World History of the Automobile

Other SAE books of interest:

**The Birth of Chrysler Corporation and
Its Engineering Legacy**
by Carl Breer
(Order No. R-144)

The Automobile: A Century of Progress
(Order No. R-203)

**Carriages Without Horses: J. Frank Duryea and
the Birth of the American Automobile Industry**
by Richard P. Scharchburg
(Order No. R-127)

For more information or to order this book, contact SAE at 400 Commonwealth Drive,
Warrendale, PA 15096-0001; phone (724) 776-4970; fax (724) 776-0790; e-mail:
publications@sae.org.

World History of the Automobile

Erik Eckermann

Translated by Peter L. Albrecht

Society of Automotive Engineers, Inc.
Warrendale, Pa.

Library of Congress Cataloging-in-Publication Data

Eckermann, Erik.
   [Vom Dampfwagen zum Auto. English]
   World history of the automobile / Erik Eckermann ; translated by
Peter L. Albrecht.
        p. cm.
   Includes bibliographical references and index.
   ISBN 0-7680-0800-X
   1. Automobiles—History. I. Title.

TL15.E282413 2001
629.222'09—dc21
2001034148

Updated English translation of
Vom Dampfwagen zum Auto by Erik Eckermann,
Copyright © 1989   Erik Eckermann

Society of Automotive Engineers, Inc.
400 Commonwealth Drive
Warrendale, PA 15096-0001 U.S.A.
Phone: (724) 776-4841
Fax: (724) 776-5760
E-mail: publications@sae.org
http://www.sae.org

ISBN 0-7680-0800-X

SAE Order No. R-272

Dedicated to my grandfather,

Willy Eckermann

to whom I owe so much.

Contents

Introduction

Thousands of years separate the invention of the wheel and the first self-propelled vehicle. The intervening centuries witnessed wind-powered vehicles (Figure 1), wheeled sailing ships, and muscle-driven vehicles (Figure 2), in which human or animal power, hidden or in plain sight, served as motive power. However, these did not represent real progress, because neither wind nor muscle power was faster, more powerful, or blessed with greater endurance than the combination of horse and wagon. In other words, development of self-propelled road vehicles depended on finding a suitable power source.

Figure 1. Wind-powered vehicle, circa 1600. A sail car built by Simon Stevin, with rear-wheel steering. The helmsman used a tiller to swing the rear axle around a kingbolt.

The availability of this power source around 1885, in the form of a lightweight combustion engine, resulted in a paradigm shift 50 years after the railway had revolutionized transportation. Suddenly, it was possible to equip not only two-wheeled vehicles, coaches, and trucks, but also ships and boats, streetcars, airships, airplanes, fire engines, and many other devices with an engine that, thanks to liquid fuel carried on board, could operate anywhere. Motorized

*Figure 2. Muscle-powered carriage, 1765. Four-wheeled muscle-powered vehicle,
treadle operated with a ratchet-and-pawl mechanism on the left rear wheel.*

road transport quickly overcame the decades-old advantage of the railroad, and set new standards for free-ranging surface transportation, time savings, and individual mobility. In the
process, mass motorization has spawned problems of its own which threaten to grow to
uncontrollable proportions: energy consumption, dwindling resources, environmental pollution,
climate changes, and traffic accidents, to name just a few. The automobile must be regarded
not only from technological or economic standpoints, but also from an ecological perspective.

The Prehistory of the Automobile

Automobile

From Ancient Times to 1884

The Development of Animal-Drawn Transportation

If we regard the wheel as their most distinguishing feature, the invention of two- and four-wheeled vehicles is lost in the mists of time. Both wheel and wagon probably arose simultaneously, in various regions of the world, around the fourth millennium B.C. This, at least, follows from inspection of the oldest representations, assuming that the objects illustrated (Figure 3) are a few centuries older. Numerous archaeological finds from around 2600 B.C., including the Standard of Ur (Figure 4), a model of a cart (a *quadriga* from Tell Agrab, Iraq), and the remains of actual carts or wheels (royal tombs of Kisch, Mesopotamia, and the peat bogs of Schleswig-Holstein) allow us to draw conclusions regarding the technology of wheeled transport in antiquity. Apparently, in the beginning, there was the simple disc-shaped wheel, made of at least two boards. These solid-disc wheels were approximately 4.5 cm (1.8 in.) thick and were fitted with "tires" made of leather strips, nails, or a copper crown. Depending on design, such a wheel weighed between 30 and 70 kg (66 and 154 lb). (For comparison, a modern 13-inch steel wheel and tire, or a 17-inch alloy wheel without tire, weighs approximately 11 kg [24 lb]).

Such wheels, usually reinforced in the hub area, turned on axles that were attached directly to the wagon floor or superstructure. There were also designs in which the wheels were firmly splined to the axle, which turned with the wheels as a unit below the floor of the vehicle (see Appendix, Wheel attachment). Due to the lack of any steering system, four-wheeled wagons tended to run straight ahead; in (gentle) turns, wear on the vehicle and the effort expended by the draft animals increased considerably.

Wagon superstructures were usually simple boxes, open to the sky. Metals, such as copper and bronze for tires, or lead and tin sheathing for individual parts, reinforced or decorated the wood that made up the boxes, wheels, axles, and wagon poles.

Figure 3. Solid-disc wheel, circa 3500 B.C. Sumerian solid-disc wheel, consisting of three disc segments, clamps, leather tire, and axle hole. Reproduction.

Figure 4. Sumerian war wagon, circa 2600 B.C. Four-wheeled war wagon, crewed by one driver and one warrior, with a front-mounted shield and quiver for javelins, pulled by four onagers (wild donkeys). Detail from the "war side" of the Standard of Ur, a wooden chest found in the royal cemetery of Ur.

A precursor of the wagon was perhaps a sled, initially rolling on logs and later on disc wheels. In this way, a "rolling" vehicle might well have evolved from a "sliding" vehicle on skids or runners. The ancestor of the sled is the sledge (Figure 5).

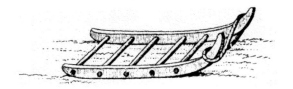

Figure 5. Sledge. Ancient Egyptian sledge with curved runners.

Both the sledge and sled are draft vehicles fitted with skids or runners. In the case of the sled (Figure 6), the sliding and load-bearing functions are separated by a high-mounted platform. Sledges and sleds rolling on logs remained in use after the invention of wheeled transport, for example, to move heavy loads for the construction of pyramids, obelisks, and temples in Egypt, around 1500 B.C. Both sledge and sled trace their origins to the travois (Figure 7), a forked branch with or without transverse members, used by primitive man to drag home the spoils of the hunt.

Figure 6. Sled. Ox-drawn sled, raised platform, parallel runners, steering by means of draft pole(s).

Development of the two-axle, four-wheeled wagon and the two-wheeled cart did not proceed at an equal pace. While the wagon served primarily to transport merchandise and was rather unwieldy and slow because of its draft oxen, lack of steering, and heavy disc wheels, the suitability of the cart as a vehicle for hunting, combat, and prestige was recognized at an early

Figure 7. Travois. The most primitive form of travois consisted of a forked branch, with or without transverse members. This constituted mankind's first land "vehicle."

stage. Around 1550 B.C., the Egyptians conducted successful campaigns against the Hittites, Syrians, and the Mittani, using chariot units as shock troops. The light chariot, rolling on spoked wheels, changed not only the art of war and order of battle, but also influenced the social and economic fabric of society.

New occupations arose, such as wagon maker, wheelwright, tool maker, and horse breeder. The arts also embraced the new artifact (Figure 8). The multitude of representations of hunting and war chariots known to us in the form of reliefs, murals, and artifacts suggests that Egyptian craftsmen and artisans were no less involved with the wheeled transport of their age than artists of the twentieth century were with the automobile.

From the Middle East, the wagon maker's skills spread to southeastern Europe. In Crete, the heavy, ox-drawn freight wagon appeared around 2000 B.C. By 1000 B.C., it had undergone its most significant development with the fitting of a pivoting front axle. The kingbolt (or

Figure 8. Egyptian war chariot, circa 1290 B.C. Lighter and more maneuverable, two-wheeled carts proved more suitable for combat and hunting than four-wheeled wagons (Figure 4). Spoked wheels have already replaced solid wheels.

4

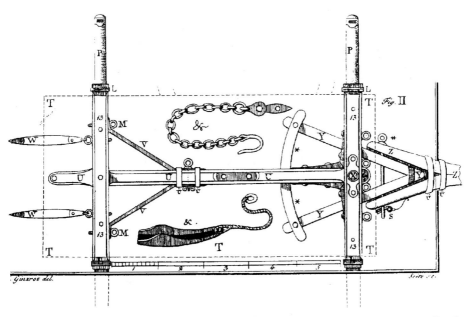

*Figure 9. Turntable steering. The yoke Z swings the arms Y, attached to the front axle P, around the kingbolt X, mounted below the front axle carrier 13. The turntable * is mounted below the perch U and keeps the front axle level.*

kingpin) pivot, presumably invented in Scandinavia, spread to southern Europe, probably through the Celts, who refined the concept first with kingbolt-and-turntable steering, then with fifth-wheel steering (Figure 9 and Appendix). These developments arrived just in time to serve the expansionist urges of the Romans, whose steerable four-wheeled wagons were, of course, not used in combat.

Light, two-wheeled war chariots were known in Mycenae from the sixteenth century B.C. Most famous was the combat, racing, and ceremonial chariot drawn by a team of horses harnessed four abreast, known as the *quadriga* by the Romans. Although four-horse teams had been approved in Greece from 680 B.C. for the Olympic Games, Roman quadriga races served as public entertainment—a precursor of modern auto racing on a permanent (circular) track. After victorious generals made a habit of entering Rome in quadrigae, its form began to grace palaces, triumphal arches, and coins beginning in the seventh century B.C.

In contrast to the Greek city-states, able to conduct their trade and colonial politics via maritime transport, the Roman Empire encompassed inland domains in Europe, Africa, and the Near East. Rome managed a road network that stretched from North Africa to Britain, and from Mesopotamia to the Iberian Peninsula. Whenever possible, these roads were laid out to join two points by the shortest route. Natural barriers were not bypassed, but rather, insofar as the technology of the time allowed, simply conquered by the most direct path.

Roman roads consisted of metalled (gravelled) roadways, approximately 3 meters (10 ft) wide. Certain roads, such as portions of the Appian Way, were up to 4.5 meters (15 ft) wide and were paved. Note that in Rome, wagon traffic was banned from the roads during the day, an early version of the pedestrian zones in modern cities.

Rome was capable of deploying troops, with or without wheeled transport, relatively quickly. In time, the war chariot declined in importance in favor of more maneuverable and more cost-effective cavalry. It was seldom used after the third century B.C. In its place, travel and commercial traffic on the roads continued to develop. Private individuals could rent two- and four-wheeled vehicles in any large town. Freight wagons were available for heavy transport.

Around the time of the birth of Christ, a state-operated courier and dispatch service was established, with waystations for changing horses. The fixed location of these stations *(posita statio)* eventually became the Italian word *posta*. Originally, the word "post" did not refer to a communications system, but rather to a permanent location for changing horses. It was not until the end of the fifteenth century that the word assumed its current meaning. Despite many crises, the Roman "postal" service proved to be so resilient that it survived the fall of the Roman Empire (478 A.D.) and was even maintained by the invading Germanic tribes for their own purposes.

Figure 10. Roman rheda. *Steerable multipurpose wagon, given various designations according to specific superstructure. Here, a horse-drawn model, with a luxurious cover, entry flap, and chain brake.*

One of the many wagon types in general use was the steerable *reda* (or *rheda*) (Figure 10), originally a Celtic vehicle, adopted and improved by the Romans. A four-wheeled flatbed vehicle with or without weather protection, the reda was suited for all transportation duties. It served as a city or long-distance carriage, postal, military, and rental vehicle, and, in a more luxurious form, even served the caesars and the most prominent citizens of the Empire "when they wished to appear without any great pomp" (Ginzrot, J.Ch., *Die Wägen und Fahrwerke der Griechen und Römer und anderer alter Völker*, 1817, p. 288).

With the fall of the Roman Empire, road traffic of that age ceased to exist. The roads fell into disrepair, and the wagon builder's art stagnated. The only means of transport, if any were available, was the horse.

It was not until the fifteenth century that road traffic began to regain its importance as central Europe experienced a general economic resurgence. The voyages of discovery (discovery of the New World in 1492 and the opening of a sea route to India in 1498) began to have an effect. They shifted the economic balance from the Mediterranean to the Atlantic and Channel coasts. Movement of wares required an expanded road network extending from the principal ports—Lisbon, Seville, and Antwerp—to the European interior.

At this time, progress in wagon making again becomes apparent. The form known today as the coach presumably originated in Hungary. This involved an enclosed compartment, suspended by leather straps that were intended to serve as springs but probably lessened only the most severe road shocks. In the seventeenth century, steel leaf springs appeared, their lower extremities attached to the chassis, with attachment points at their upper ends for leather straps carrying the bed or cabin.

The diameter of the rear wheels increased, as larger wheels ran more smoothly on uneven roads. This raised the level of the bed, which improved the action of springs and straps (Figure 11). To improve maneuverability, smaller front wheels were installed, allowing them to turn under the front structure. Instead of a flat disc, the wheels were given a "dished" shape, necessitated by the camber of the axle stubs (Figure 12), which for safety reasons had the hub turning against a thrust washer instead of the axle nut. Because the lowermost spoke must be perpendicular to the road to avoid collapse, the spokes also were angled with respect to the hub (spoke camber, Figure 13).

By contrast, freight vehicles changed little or not at all. Agricultural wagons or the heavy freight wagons of the merchants, which rumbled their way from town to town and across Europe, remained unsprung, unwieldy, and in many cases unbraked. Except for introduction of the fifth wheel, freight vehicles had undergone virtually no improvement since antiquity.

This is surprising in view of the fact that freight vehicles were needed in all eras, even though the passenger wagon had been replaced by the horse in medieval times and did not return to fashion until the beginning of the Renaissance. The German technologist and economist Johann Beckmann, writing in *Beyträgen zur Geschichte der Erfindungen* ("Contributions to the History of Invention," 1782) writes that "even ladies found it difficult to use covered transportation, and gentlemen regarded it as unseemly to ride in vehicles."

Figure 11. Coupe, circa 1810. The carriage body is suspended by steel C-springs and leather straps. A perch joins the front and rear axles. For turning or reversing, the small front wheels are able to pass below the front structure. Dick & Kirschten, Offenbach.

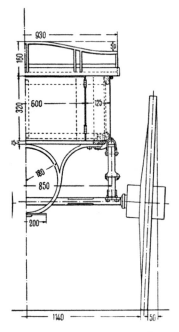

Figure 12. Axle camber. To prevent the wheels from running off the axles, the axle stub is given a downward camber. As a result, the wheel hub runs against the axle flange (thrust washer) instead of the axle nut.

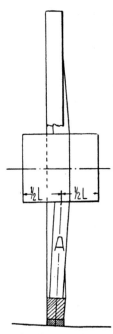

Figure 13. Spoke camber. With cambered axles, it is also necessary to camber the wheel spokes. To carry wheel loads without breaking, the lower spoke must be perpendicular to the ground.

A change in attitudes occurred in Spain during the sixteenth century. At that time, Spain, by virtue of its political and economic importance, established the lifestyle and fashion of virtually all European courts. Nobility accepted the use of horse-drawn carriages. In the process, ever more ornate coaches were made, becoming not so much utilitarian devices as status symbols.

The catastrophic demise of the Spanish Armada in 1588 announced the ascendancy of England as a political, economic, and technological superpower. Unlike Spain, the use of coaches in England was not limited to the court and nobility, but found favor among the bourgeoisie. Countless light, capable carriage and wagon designs arose. From the eighteenth century, virtually all improvements in coachbuilding originated in England, where travel by means of post chaise, post chariot, or stage coach came into widespread use.

The invention of leaf or elliptical springs by the English builder Obadiah Elliott in 1805, replacing the previously typical leather straps and elbow or C-shaped springs (Figure 14), represented a significant advancement. Their installation between the axles and box eliminated the need for the perch (longitudinal member) stretching between the front and rear axles. This reduced weight and step-up height, and improved springing comfort and road manners. The elimination of the

Figure 14. Coupe, circa 1890. The carriage body is suspended by full elliptic springs front and rear; the perch is eliminated (self-supporting coachwork). This makes the carriage lighter and cheaper to produce. Mayer'sche Wagenfabrik, Bayreuth.

perch as the fundamental chassis characteristic also represented the transition to self-supporting coachwork, which today is customary on passenger cars (in combination with more modern suspension systems). In commercial vehicles, however, ladder frames remain the state of the art, tracing their origin to the double perch or brancards of Berlin-type coaches.

Braking systems were known from the early seventeenth century. To assist the braking action of draft horses and the efforts of the wagon crew, who braced themselves against the wheel spokes, brake shoes were introduced—U-shaped forged iron channels that were wedged under the rear wheels. In this way, wheel rotation could be halted instantly, and the resulting friction between the wheel and roadway, with its rather limited braking effect, increased rolling resistance. Modulated braking was introduced in the mid-nineteenth century in the form of linkage-actuated brake blocks, still found on freight wagons today.

After 1850, iron replaced wooden axles. A notable improvement was achieved with permanent oil lubrication of the hubs and adjustment for axial play of the wheels. Now, the wheels no longer needed to be removed from the axles for periodic lubrication and adjustment. These patented axles, by John Besant (1786) and John Collinge (1787), also originated in England.

The Age of Enlightenment saw the first scientific study of wagon building, "nearly 5,000 years after the invention of the wagon and about 100 years before that of the steam locomotive" (Treue, W., *Achse, Rad und Wagen* ("Axle, Wheel and Wagon"), Munich, 1965, p. 226). Beginning in 1717, academies in Stockholm, Paris, and Copenhagen encouraged research into the scientific theory of wagon building. The following decades saw the publication of dissertations, studies, and books on the topic. The sections of the French Encyclopedia of Denis Diderot and Jean le Rond d'Alembert (1772) covering the art and technology of wagon building and harness making were exemplary, as was the work of Beckmann in Germany. The improvements suggested in these writings, such as taller wheels, thinner iron axles, and oil instead of pine tar lubrication, had a beneficial effect on wagon-building technology and simultaneously represented the birth of a scientific basis for vehicle technology. The wagon-making industry and its craftsmen, artisans, and stylists, as well as blacksmiths, harness makers, trimmers, and painters, raised the craft to a high level.

The Quest for a Prime Mover

The wagon existed in its animal-drawn form for thousands of years before it was possible to make it self-propelled, literally "auto-mobile." In the process, motorized vehicles were far removed from the center of scientific and technical inquiry. From the end of the seventeenth century, existing vehicular technology was more than adequate to meet societal demands. In the age of absolute monarchs and mercantilism, it was more important to solve other engineering challenges that were difficult or impossible to achieve with conventional energy sources such as muscle, wind, or water power. The fountains and water displays of baroque gardens, and especially pumping water out of mines, occupied the attention of many learned men.

The Dutch physicist, mathematician, and astronomer Christian Huygens (1629–1695) applied gunpowder as an energy source. In 1673, building on the experiments of German naturalist and statesman Otto von Guericke (1602–1682), Huygens used the explosion of gunpowder in a cylinder to achieve partial vacuum. Atmospheric pressure drove the piston back down the cylinder, resulting in work (Figure 15). Although experiments and demonstrations with the small gunpowder engine were satisfactory, and Huygens immediately turned his thoughts to the invention of "new types of vehicles for water and land..." (Klemm, Friedrich, *Kurze Geschichte der Technik*, "Short History of Technology," Freiburg, 1961, p. 107), it was not possible to achieve a continuous cycle with this machine. Nevertheless, Huygens' gunpowder engine is the oldest device known to employ combustion in a cylinder for the release of energy. It may be considered the predecessor of the atmospheric gas engine.

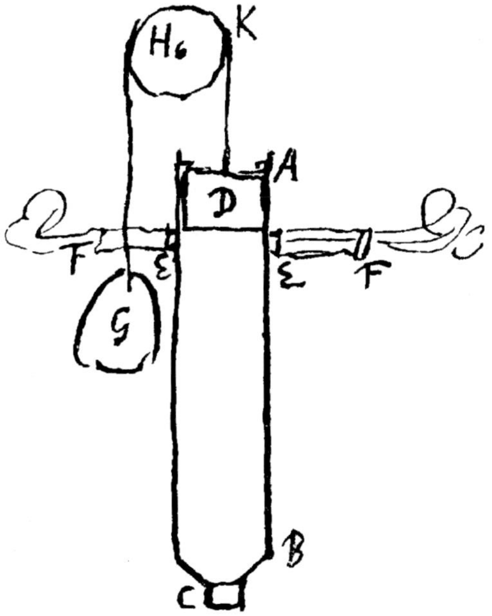

In France, Denis Papin (1647–1712 or 1714), despite several improvements exhibited by his own gunpowder engine of 1688, which had Huygens' apparatus as its inspiration, was unable to achieve satisfactory operation. The machine proved to be unsuitable for further development, if for no other reason than the danger posed by its fuel source, and did not encourage any other inventors.

In 1690, Papin built a new machine in which the cylinder was evacuated by the condensation of steam. It

Figure 15. Huygens' gunpowder engine, 1673. Gunpowder is ignited in pan C. The resulting hot gases drive piston D upward in cylinder A-B, and escape through leather hoses E-F, which act as valves. Atmospheric pressure equalizes the partial vacuum inside the cylinder by driving the piston downward and closing the leather hoses. In the process, load G is lifted by a rope and pulley H.

was not suitable for practical applications but exhibited the basic principles of the later atmospheric steam engine.

In the years that followed, it would appear that every inventor, natural philosopher, and mechanic devoted his attention to steam engines (see Appendix). In 1698, Thomas Savery of England built a steam pump that alternately used the expansive power of steam and atmospheric pressure. It was able to pump water by means of a lever mechanism; however, because of the large head required in mines, it was at first of little use.

In 1712, Thomas Newcomen (1663–1729) built a more effective engine. He separated the steam boiler from the working cylinder, and used water to cool the steam in the cylinder, allowing atmospheric pressure to push the piston back into the resulting partial vacuum. To lift water out of mines, Newcomen connected the piston to the pump linkage by means of chains and a rocking beam.

Although Newcomen's atmospheric steam engine found wide application in England, it was rather inefficient. The cylinder had to be as hot as possible when steam was admitted, but as cold as possible during condensation to achieve good partial vacuum. However, an alternately hot and cold cylinder implies large heat and energy losses.

James Watt (1736–1819) of Scotland recognized these physical processes and introduced practical improvements. In 1768–1769, Watt built an experimental engine using a working cylinder that remained hot, and a separate, continually cold condenser. Figure 16 shows that not atmospheric pressure, but rather steam itself, was its driving force. This then represented the invention of the direct-acting steam engine. The single-acting steam engine manufactured by Boulton and Watt beginning in 1776 consumed only half as much coal as an improved Newcomen engine. Altering steam pressure even allowed the power output, which amounted to approximately 20 horsepower, to be modulated.

Initially, the Cornish mines, fighting an ever more difficult battle against water seepage, were the main customers for Watt's steam engine. Soon, however, other burgeoning industries, especially the iron industry and textile mills, demanded machines with rotary power output.

Patent considerations prohibited Watt from employing the crank (see Appendix) as a mechanical component to convert reciprocating motion into rotary motion. In 1781, he applied for a patent for the planetary gear system (see Appendix) made popular by his machines. Its actual inventor, however, was his assistant William Murdock (1754–1839).

An additional, vital improvement was Watt's double-acting steam engine of 1782. To transfer power from the connecting rod, which gave power on both strokes, to the walking beam, he developed a parallel linkage (the Watt mechanism) to replace the earlier chain linkage, which worked in only one direction. To control engine speed under changing load, he invented the centrifugal governor.

Watt's steam engine was the catalyst for a wave of rapid mechanization and industrialization, indeed for the entire Industrial Revolution, which originated in England. When used to drive water pumps, the steam engine allowed sinking coal and ore mines to greater depths than ever in the past. In turn, its manufacture required other machines, leading to the development of new machine tools. The steam engine led to the creation of entirely new industries and to a growing demand for coal and iron.

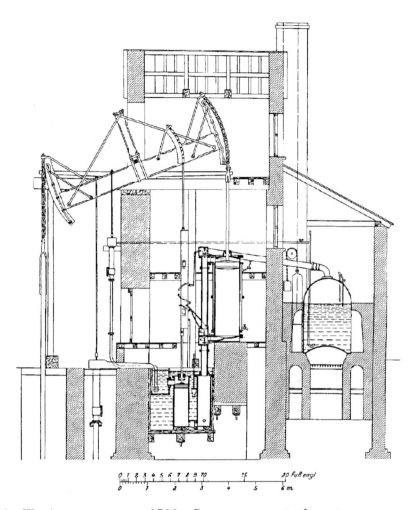

Figure 16. Watt's steam engine, 1788. Reciprocating single-acting steam engine used to drive a water pump, to British patent 913 dated January 5 and April 25, 1769. Operation: the space above the piston in the working cylinder (center of illustration) is continuously connected to the boiler (building at right side). With the piston at the bottom of its travel, steam is admitted below the piston. The same steam pressure exists on both sides of the piston, and the weight of the pump linkage (left) pulls the piston upward. At this point, the space below the piston is connected to the condenser (at left, below the working cylinder). Steam below the piston condenses, and the higher steam pressure above the piston drives it back down again, resulting in work being performed through the balance beam.

The steam engine opened new horizons in agriculture and transportation. The steam plow blazed a trail for the diesel tractors in worldwide use today. The steam railway locomotive gave mobility to the masses on an unprecedented scale. Steamships transported millions of emigrants to distant lands.

An obvious improvement was to make the heavy, stationary steam engine smaller and lighter, and apply it to powering a vehicle. In 1769, at the instigation of Etienne-Francois Choiseul-Ambroise, the French foreign minister, engineer Nicolas Joseph Cugnot (1725–1804) became the first to design and build a steam-powered vehicle—a three-wheeled conveyance capable of carrying four persons. After several satisfactory test drives, it was followed in 1771 by an improved heavy freight wagon (*fardier*) of the same general layout (Figure 17).

Figure 17. Cugnot's steam wagon, 1771. After the fardier *was built, the new political masters of France had no more interest in Cugnot's steam wagon. Today, it is preserved in the Paris Conservatory. As the story goes, several officials took a test drive, which ended with the first motor vehicle accident.*

Cugnot abandoned the customary atmospheric principle of the time and instead employed steam under pressure, eliminating the condensation arrangement and its controls. He grouped the boiler, engine, and drive around the single front wheel, giving a large cargo platform for the transport of munitions. This tractor (for towing artillery) required a steady flow of power, which Cugnot provided with two cylinders, alternating power strokes, and a logical coupling of the connecting rods by means of a rocker arm and chain links. Cugnot converted reciprocating motion to rotary motion by means of a ratchet and pawl mechanism, which also permitted operation in reverse (Figure 18).

An official demonstration was never performed because the project's sponsor, Choiseul-Ambroise, fell out of court favor in 1770, in the chaos preceding the dissolution of the French monarchy. His successors apparently saw no need for a mechanically driven military vehicle. Cugnot's freight wagon was the first mechanically propelled, functional road vehicle in history. It is preserved in the Conservatoire National des Arts et Métiers, in Paris. A full-size replica has been built, as well as several scale models, of which one (one-fifth scale) is on display at the Deutsches Museum in Munich.

In England, Richard Trevithick's experiments with high-pressure steam engines led to the construction of a steam carriage, which in 1803 was employed with some success in London for the transport of passengers—the first motorized taxi (Figure 19). However, satisfactory conditions for operation of steam omnibuses were not met until approximately 20 years later—again in England, whose technological and industrial superiority was unchallenged at the time. Sir Goldsworthy Gurney built several steam omnibuses (Figure 20), one of which covered the 14 km (8.7 miles) between Gloucester and Cheltenham three times daily. Its average speed, including stops, amounted to approximately 18 km/h (12 mph). In 1831, Walter Hancock established regular daily bus service between London and Stratford with his Infant, capable of carrying 14 persons.

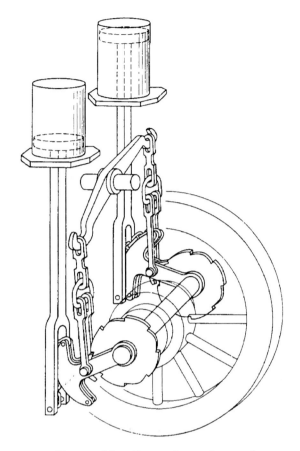

Figure 18. Cugnot's ratchet-and-pawl drive, 1771. Cugnot achieved the continuous power flow required by a vehicle by applying alternating power strokes from a two-cylinder engine. Piston rods, bellcrank, chains, and linkages transmitted power to a ratchet-and-pawl mechanism, which converted the reciprocating motion of the pistons into rotary motion.

It was apparent that the technology was not yet fully developed, and this new means of transportation did not yet enjoy favorable public opinion. Crankshafts snapped, lines leaked, chains broke, and boilers exploded. Engine vibrations (which, unlike stationary installations,

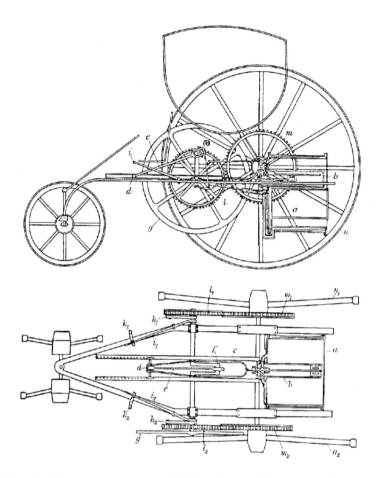

Figure 19. Trevithick's steam carriage, 1803. Patent drawing indicating a passenger-carrying body suspended high above the drive. a, *boiler;* b, *cylinder;* c, *forked piston rod;* e, *connecting rod;* g, *flywheel. A working reproduction was built in 1998.*

could not be overcome by mounting on a solid foundation), the pungent odor of burnt oil, and flying soot and coal dust soon drove the traveling public back to the old standby, the horse-drawn stage, or another new invention, the railway and its rapidly growing network of track.

Steam omnibuses charged higher fares than horse-drawn buses. The 9-km (6-mile) distance between Liverpool and Prescott cost 48 shillings for steam transport, but only 4 shillings for the horse-drawn omnibus. The public was unappreciative of the new contraption; indeed, it took a hostile attitude. Investment capital shifted to the railways, and engineers turned their attention to the locomotive and steamship. Around 1840, development of steam omnibuses came to a virtual standstill in England.

The few engineers and companies that continued to occupy themselves with mobile steam technology concentrated on portable steam engines (horse-drawn wagons with mounted steam

Figure 20. Gurney's steam omnibus, 1828. Around 1830, steam omnibuses appeared in England, achieving average speeds as high as 20 km/h (12 mph). Shown here is one of Gurney's steam coaches. Unknown to the inventor, its two additional pilot wheels actually worked against his fifth-wheel steering geometry, as the extensions of the three axles did not meet at a single point when taking a bend.

engines for agricultural purposes) and traction engines (low-speed steam tractors for towing heavy loads, powering agricultural machinery, and drayage). With their large, horizontal boilers, they more closely resembled roadable locomotives and cannot truly be considered forerunners of commercial road vehicles.

Traction engines proved to be a great asset to agriculture; their numbers increased steadily until 1865. Some were exported, and in Germany they were called "locomobiles." In England, however, traction engines drew the enmity of railway inspectors, owners of freight companies, and "gentlemen drivers" because they damaged roads and bridges and because they frightened horses.

Bowing to public pressure, Parliament passed a law in 1865 which, among other things, limited top speed to 6.4 km/h (4 mph) outside towns and 3.2 km/h (2 mph) within townships, set weight and size limits, and ordained that every steam vehicle traveling on public roadways must be preceded by a person carrying a red flag, to warn horse and human alike.

This so-called Red Flag Act had disastrous consequences for British industry. Development and manufacture of vehicles powered by steam or other mechanical means came to a complete halt.

The lessons learned in building lightweight vehicles such as those of Hancock and Gurney were lost. Incentive for more intensive engineering activities and capital investment did not return until 31 years later when the law was repealed in 1896.

By that time, it was too late for England. After the Franco-Prussian War of 1870–1871, France had taken the lead in steam vehicle construction. French engineers also established the theoretical basis for the predecessors of our modern combustion engines. The first motorcars appeared in Germany during the mid-1880s. The first gasoline-powered English cars did not appear until nearly a decade later—Bremer in 1894, and Knight in 1895—while the Continent already boasted about ten motorcar manufacturers and Benz was producing its Velo model in a limited series. With the Red Flag Act, the British Parliament had turned the "workshop of the world" into a back-alley tinkerer's shed.

In France, Amédée Bollée the elder concluded the experimental stage of his steam omnibus experiments in 1873. Albert de Dion (1856–1946) began his steam vehicle research in 1883, concentrating on reducing the extreme mass and the resulting lack of maneuverability. He adopted bicycle construction methods, including tubular frames, wire spoke wheels, and steering forks. Léon Serpollet also was an advocate of lightweight design, beginning in 1887. He made numerous improvements to steam engine technology, such as reducing the time needed to raise steam (*vaporisation instantanée*—his so-called flash boiler) and automatic coke firing.

By this time, steam vehicle design had reached a high level of technical development. However, the conditions were not yet ripe for motorizing a broad segment of the population, for reasons both technical (e.g., heavy construction and complicated operation) and social (e.g., limited buying power and no real need). The steam engine made possible a continuous working cycle, but its thermal efficiency (see Appendix) of approximately 10 percent (1897) was inadequate. It remained for the gasoline engine, whose development history begins with gas engines, to open completely new possibilities for road, rail, water, and air transportation.

At the end of the eighteenth century, the invention of coal gas, produced by the decomposition of coal and other fuels by heat in the absence of air, spurred considerations for design of gas engines. As the name indicates, these are combustion engines powered by fuels in gaseous form, in which the mixing of flammable gases and air usually is accomplished in a separate chamber outside the working cylinder.

Around 1800, Francois Isaac de Rivaz (1752–1829) of the Valais (a former republic then under Napoleonic rule, and today a Swiss canton) experimented with an engine employing an open cylinder in which a mixture of hydrogen and air was ignited electrically. The piston, flung upward by combustion pressure and returned by atmospheric pressure, operated a ratchet mechanism to motivate a four-wheeled platform (Figure 21) for a distance of a few meters, after which fresh mixture would have to be introduced manually. Test runs were conducted in 1809 and 1813; the French patent is dated 1807. The work of de Rivaz, who had built steam cars before his gas engine and continued to do so afterward, represents the first attempt to harness an internal combustion engine to drive a road vehicle.

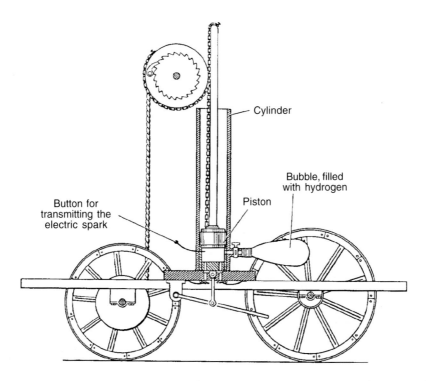

Cylinder

Bubble, filled
with hydrogen

Button for
transmitting the
electric spark

Piston

Figure 21. de Rivaz' vehicle, 1809 and 1813. Completed patent drawing of 1807. An electrically ignited hydrogen charge hurled a piston up the open cylinder. Falling back down, a chain attached to the piston rod turned a pulley (fitted with a ratchet), which in turn drove the rear wheels by means of a rope and second pulley. In 1984, the Vevey Vocational Training Center built a one-third-scale working replica of de Rivaz' vehicle.

Jean Joseph Etienne Lenoir of Luxembourg (1822–1900) finally succeeded in lifting gas engine technology out of the experimental stage. His gas engine was fully operational, even if it was uneconomical. Lenoir's double-acting engine had sliding inlet and exhaust valves and resembled the horizontal steam engines of his day, both technically and visually (Figure 22).

Beginning in 1860, several Paris workshops built Lenoir's gas engine under license, the first example of an engine being built on an industrial scale (300 to 400 examples). Its disadvantages were its excessive consumption of lubricating oil and gas, traceable to its lack of precompression, and its rough running characteristics, caused by ignition at mid-stroke, when piston speed is greatest. As such, it offered small operators a (barely) viable alternative to the complexity of the steam engine, which could be employed economically only in large plants.

As an experiment, Lenoir installed his gas engine in a modified break (Figure 23); however, this proved to be an unsuitable power source for a road vehicle because of the need for the vehicle to carry its own gas supply.

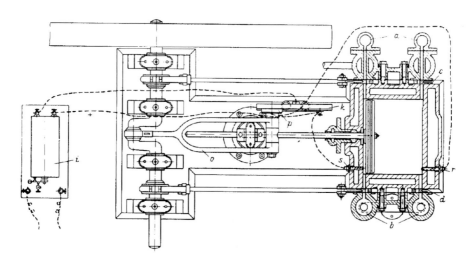

Figure 22. Lenoir's gas engine, 1860. Double-acting two-stroke with gas admitted by a slide valve c; *gas and air are mixed in the valve body and adjoining passage. With piston moving to the right, the mixture is ignited by spark plug* s *at about mid-stroke, accompanied by a sudden rise in pressure (with attendant noise and shock loading) and acceleration of the piston, with energy transfer through connecting rod* o *and crankshaft (with flywheel). Exhaust gas escapes through the exhaust slide valve* d *on the left side, into exhaust pipes* b. *The process is repeated on the right side (i.e., double acting), to be ignited by spark plug* r. *The induction device is at* i, *with ignition distributor* k *and contact disc* p.

As a stationary power source, however, the Lenoir engine gained widespread acclaim throughout Europe. This prompted Nikolaus August Otto (1832–1891), a businessman in Cologne, to improve upon the French engine. Above all, he wanted to adapt it to liquid fuels such as alcohol, to gain independence from town gas. He even considered a machine "...for locomotion of vehicles on country roads..." (patent application of 1861).

Experiments with a replica Lenoir engine led Otto to realize that the gas/air mixture had to be compressed inside the cylinder and ignited at the moment of maximum pressure—that is, at top dead center, just before it began its downward motion. An experimental four-cylinder engine of his own design, built in 1862, was destroyed by its own powerful firing pulses. Dejected, Otto terminated his experiments and took up the construction of atmospheric gas engines.

Even then, atmospheric gas engines did not enjoy any development potential, except in terms of their height which, however, made them unsuitable for vehicular power. Declining sales forced the firm of N.A. Otto & Cie, and its successor the Deutz Gas Engine Works AG (Gasmotoren-Fabrik Deutz AG) to develop a more powerful engine of new design.

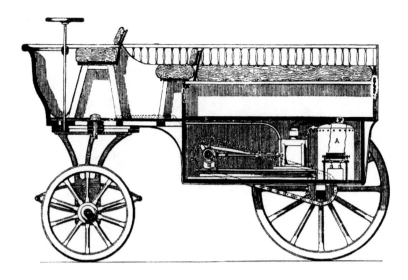

Figure 23. Lenoir's gas engine vehicle, 1863. Lenoir allegedly made several runs with this three-wheeled break in 1862 and 1863. Assuming that A is the gas tank, its range could not have been great. Also, there is no indication of a flywheel.

To that end, in 1876 Otto resumed the Lenoir gas engine experiments he had terminated fourteen years earlier. This time, he employed a single-cylinder test engine which, probably instinctively or perhaps "with a stroke of genius" as it were (Sass, Friedrich, *Geschichte des Deutschen Verbrennungsmotorenbaues,* "History of German Combustion Engine Design," Berlin, 1962, p. 43), operated on a four-stroke cycle. The gas-fired engine immediately developed as much power as the largest atmospheric gas engine the firm had been building since 1863—namely, all of 3 hp.

Otto had designed his engine to be single-acting and in principle employed the familiar flame ignition of atmospheric engines (see Appendix, Ignition systems). For speed regulation, he employed a "hit or miss" control that shut off the gas supply when the engine was unloaded and oversped.

In the same year, Wilhelm Maybach (1846–1929), who at this time was employed by Deutz, built a somewhat larger machine (Figure 24). He gave it a more pleasing appearance, replaced the hit-or-miss control with a Watt-like centrifugal governor, and also employed a crosshead—a guided intermediate link joining the pure reciprocating piston rod to a connecting rod to transfer force to the crankshaft.

In this form, Otto's four-stroke engine of 1876 ended the 200-year-long search for an engine suited to the needs of craftsmen, small shops, and, ten years later, road vehicles.

Figure 24. Otto's gas engine, 1876. Single-acting four-stroke without crosshead ("plunger piston"), with hand control for gas supply. Slight improvements by Maybach in the same year (crosshead and centrifugal governor) and made to visually resemble the English horizontal steam engines of the time. Single-cylinder, 6104 cc, 3 hp (2.2 kW).

The priorities outlined by Otto in his German Reich patent number 532, dated August 4, 1877, require some explanation. For Otto, the working process of his engine (i.e., two- or four-stroke) was apparently of little significance because he did not seek any patent protection for the four-stroke cycle. His main goal was to reduce the impact on the piston of the powerful pressure pulses resulting from ignition of the mixture.

Otto attempted to achieve this by means of stratified charge. A rich mixture near the ignition source ensured firing, with an increasingly lean mixture closer to the piston. Patent number 532 is worded accordingly. The first three of the five claims in the patent concern mixture formation and ignition; the four-stroke cycle is not mentioned until the fourth point. Because Otto was so deeply convinced of the overwhelming importance of stratified charge, he and the Deutz gas engine company litigated against such firms building two-stroke engines which he felt violated the first three claims of his patent.

However, in its verdict of January 30, 1886, the German Imperial Supreme Court declared claims one through three as void, and rightfully so. Today, we know that smooth combustion may be achieved by other means—ignition timing, fuel quality, and combustion chamber shape, for example, but not by stratified charge. The fourth claim, the four-stroke cycle, also did not survive legal review. French engineer Alphonse-Eugène Beau de Rochas had elucidated the four-stroke principle in 1861, but without realizing its importance and, most importantly, without building a corresponding working engine. Thus, today Otto is considered the inventor of the four-stroke cycle, although he did not consider it a significant element of his engine (see Appendix, Four-stroke cycle).

The Pioneering Era and Coming of Age

1885 to 1918

The Four-Stroke Engine and Early Vehicles by Daimler, Maybach, and Benz

The 1886 German Imperial Supreme Court verdict against Otto allowed other firms to build two- and four-stroke engines (see Appendix). Now, nothing stood in the way of rapid growth for this new motor industry. Even before patent number 532 was struck down, men of vision were advancing the concept of automotive engines.

One of these was Conrad Krauss, vice president of the Hannoversche Maschinenbau AG (Hannover Machine Works—Hanomag). Krauss regarded motorization of rail traffic as both desirable and necessary; in 1877, under his leadership, engineer Georg Lieckfeld modified a two-stroke opposed-piston engine patented by Ferdinand Kindermann, the first such design in history. Converted to four-stroke operation, the Kindermann-Lieckfeld opposed-piston engine was intended for a "gas-powered locomotive," fuelled by town gas. Krauss received a patent (DRP 6768) in 1878, nine years before Daimler's Draisine of 1887, which already sported a gasoline engine.

In his patent, Krauss also covered underfloor engine mounting, horizontal single-cylinder engines, idle control, friction clutch, cam-actuated intake valves, reversing transmission, foot-operated clutch and brake, and four-wheel drive. In some cases, his proposals were decades ahead of his time. Even if no vehicles were built to his patent, Krauss nevertheless may be regarded as the father of motorized rail transport.

Krauss' innovations did not end there. As early as 1880, Lieckfeld installed a single-cylinder two-stroke Wittig & Hees gasoline engine in a railway chassis (Figure 25) and conducted early test runs. With this device, the Hannover Machine Works created the world's first gasoline-powered vehicle and demonstrated its effectiveness—five and seven years before Gottlieb Daimler and Wilhelm Maybach with their respective motorcycle and motor carriage, and six

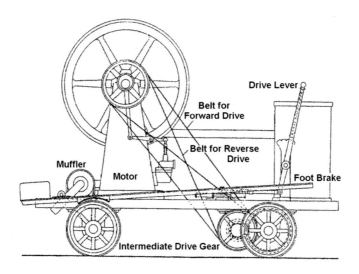

Figure 25. Hanomag gasoline rail vehicle, 1880. In 1880, the firm that would later be known as Hanomag built the world's first gasoline-powered vehicle, designed by Georg Lieckfeld and years ahead of Benz and Daimler/Maybach. However, this was not a road vehicle but rather a stationary gas engine converted to gasoline and mounted on the chassis of a streetcar drive unit. The engine employed a spray carburetor, but with its weight of approximately one ton, its high-mounted crankshaft, and its considerable size, the engine would have been unsuitable for a road vehicle. Single-cylinder engine, approximately 3 hp (2.2 kW).

years before Carl Benz and his patent motorcar. However, note that Daimler, Maybach, and Benz installed their gasoline engines in road vehicles, and the size and mass of the Hanomag engine made it totally unsuitable for road vehicles.

In Lieckfeld, the (gasoline) engine industry has yet another inventor, even if he personally did not grapple with the problems of reducing the size of stationary engines. Nevertheless, miniaturization of existing machines was a prerequisite for applying internal combustion to all types of vehicles. Although Daimler conceived applications not only in rail and road vehicles but also in boats, balloons, and fire engines, and on a grander scale considered the craft and business of the engine, Benz set his sights on creating a motorcar that unified engine and vehicle. Achieving a lighter engine design with adequate power was possible only with higher engine speed, which in turn demanded far-reaching engineering changes to existing stationary engine design.

In 1882, Gottlieb Daimler (1834–1900) and his collaborator Wilhelm Maybach (1846–1929), both of whom had been in leading positions with Deutz since 1872, established a small experimental laboratory in Cannstatt, near Stuttgart. Based on their experiences at Deutz, both men were convinced that only a four-stroke engine merited development. Because it had only half as many power strokes as a two-stroke engine in any given time interval, they expected fewer problems. At the time, the four-stroke concept remained protected by patent number 532; therefore, Daimler and Maybach had to work in secret.

Flame ignition, a proven technology on atmospheric gas engines (turning at approximately 150 rpm) and barely adequate for the requirements of stationary Otto-cycle four-stroke engines (running up to 200 rpm), was unsuitable for their engine concept. Flame ignition was ruled out by their target speed of 600 to 800 rpm, its requirement for massive slide valves and inherent sealing problems. Maybach settled on hot-tube ignition (see Appendix, Ignition systems) based on an 1881 patent by an Englishman named Watson.

At this time, there were not yet any firms specializing in ignition, and ignition devices were not available "off the shelf." Each engine constructor had to devise his own solution. This led to the development of diverse ignition systems, all arriving on the scene in rapid succession.

No less decisive for mobile application of the internal combustion engine was the switch from town gas to liquid fuel, which was more easily carried in a vehicle. The alternative was an on-board gas generating plant with concomitant weight and space requirements (e.g., producer gas generators) or a gas storage container (e.g., Lenoir, 1863).

Beginning in 1875, while still at Deutz, Maybach had experimented with devices that could atomize volatile gasoline and mix it with air. These devices, today still incorrectly labeled "carburetors," demanded constant attention to fuel level. Maybach improved these so-called surface carburetors by installing floats (see Appendix, Float carburetor) which maintained a steady fuel level and thereby a constant air/fuel ratio, and provided combustible mixture even if the vehicle was not on level ground. The float principle may be found on almost all modern carburetors, insofar as carburetors are to be found at all.

Parallel to their development of ignition and carburetor, Maybach and Daimler worked on making a more compact engine better suited for installation in vehicles. Otto's stationary four-stroke engine, weighing approximately 1500 kg (3307 lb), was out of the question. Size and weight had to be reduced, while maintaining or indeed increasing power output.

In 1884, after several intermediate stages, Maybach achieved this often overlooked but nevertheless exceptional technical achievement with his so-called "grandfather clock" engine (in German, *Standuhr*—a freestanding clock). This engine, with its vertical cylinder mounted on a barrel-shaped crankcase, displaced 462 cc and developed 1 hp at 600 rpm. To pare weight, Maybach discarded conventional crosshead design in favor of the now familiar piston and single connecting rod arrangement, made the housing walls as thin as possible, and used air cooling instead of water cooling.

A year later, with a somewhat less powerful version of the "grandfather clock," Daimler and Maybach had for the first time a small, lightweight engine at their disposal (Figure 26). Instead

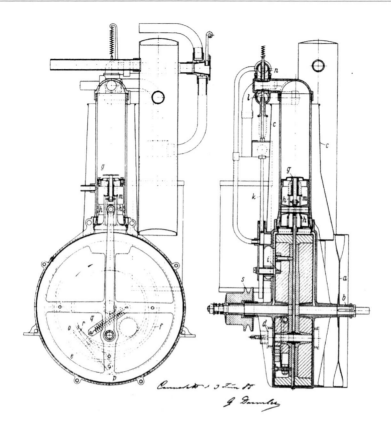

Figure 26. Gasoline engine for the Daimler/Maybach two-wheeler, 1885. A masterful accomplishment by Maybach, whose prior experience was in heavy engines: a gracefully designed, lightweight engine for a two-wheeled vehicle, with 4- and 2.5-mm (0.16- and 0.1-in.) thick walls for working cylinder and crankcase, respectively. The fan a, *configured as an axial blower, forces cooling air into the shroud* c. *Single-cylinder, 212 cc, approximately 0.5 hp (0.37 kW).*

of a four-wheeled car, their testbed was a two-wheeler, which, although not especially suited for a revolutionary new mechanical propulsion system, did represent a significant leap in the history of engine development.

This motorized cycle placed greater demands on weight reduction, compact design, and technical detail solutions than would a motorcar engine. Maybach's engineering achievement is all the more impressive when you consider that he progressed from conceiving heavy, stationary gas engines at Deutz to designing a two-wheeled powerplant whose weight of approximately 50 kg (110 lb) was only fractionally that of the Otto engine but developed almost five times its specific output.

Maybach located the engine in the frame of a two-wheeler (Figure 27)—also of his own design—precisely where it remains to this day, between the wheels and below the saddle. In

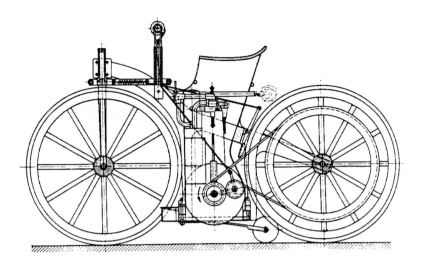

Figure 27. Daimler/Maybach two-wheeler, 1885 (patent drawing). Maybach presumably based his two-wheeler on an 1817 "walking machine" designed by Drais; both vehicles employed a wooden frame. In November 1885, Maybach undertook his first excursion with this vehicle, covering a distance of 3 km (1.9 miles) without any major problems. The actual two-wheeler is preserved in the Deutsches Museum in Munich.

late 1885, after several successful test runs, it seemed that a powerplant suitable for road vehicles had finally been born.

The two-wheeler was followed in March 1887 by the "motor carriage" (Figure 28), an open carriage with fifth-wheel steering and running gear slightly modified to allow installation of the new mechanicals. The single-cylinder engine, of 1413 cc and 1.5 hp (1887), was an up-rated version of the two-wheeler engine. A belt drove a countershaft mounted parallel to the rear axle; gears on the countershaft in turn meshed with larger gears attached to the spokes of both rear wheels. Although the engine towered above the floorboards between the bench seats in a rather ungainly and inorganic fashion, this nevertheless represented the world's first practical four-wheeled motorcar.

Around the same time, in Mannheim, not far from Cannstatt, Carl Benz (1844–1929) succeeded in building his own road vehicle, which also was powered by a gasoline engine. As co-owner of the Rheinische Gasmotorenfabrik Benz & Co., builder of two-stroke gas engines, Benz faced the same challenges as Daimler and Maybach—namely, weight and size reduction. Benz likewise chose a four-stroke cycle for his motorcar because he saw no possibility of simplifying the complexity of the two-stroke engine.

Although his horizontal engine bears an outward resemblance to the gas engines of its day (Figure 29), Benz, unable to draw on the technical genius of a Maybach, nevertheless achieved remarkable breakthroughs.

Figure 28. Daimler/Maybach motor carriage, 1887. In 1886, Daimler had the Maschinenfabrik Esslingen (Esslingen Machine Works) install an engine derived from the two-wheeler in an open carriage. The angled "skirt" below the rear seat is a radiator installed in 1887 for the now water-cooled engine. Single-cylinder engine, 1413 cc, 1.5 hp (1.1 kW) (data for the 1887 version).

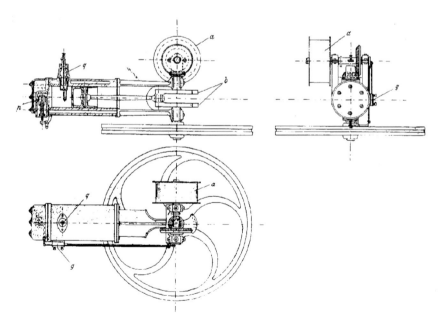

Figure 29. Benz four-stroke engine, 1884. The engine later installed in the 1885/1886 motorcar is not entirely represented by this drawing. Pulley a, crank cheeks b acting as counterweights, inlet slide valve g, exhaust valve p, spark plug q.

One of these breakthroughs was doubtless the challenge of ignition, which Benz himself described as "the problem of problems." As early as 1882, he had developed Lenoir's trembler coil ignition (see Appendix, Ignition systems) to a high level. In 1885, he again improved it for his four-stroke engine. His timed vibrator ignition was superior to Maybach's untimed hot-tube ignition, in that it permitted higher engine speeds. Admittedly, Benz' first four-stroke engine, which turned 400 rpm, did not take advantage of this feature.

Similar to Maybach, Benz also was forced to use a float-equipped surface carburetor (see Appendix), which was fitted with a preheat arrangement. A conspicuous feature of Benz' engine was its horizontal flywheel. Benz feared that gyroscopic forces on an upright wheel might compromise the stability of his three-wheeled motorcar in curves (Figure 30). Of course, the gyroscopic moment would have been far too small to make a difference. Later Benz vehicles used vertical flywheels.

On July 3, 1886, Carl Benz undertook his first trials with the "patent motorcar" on Mannheim's Ringstrasse. For the first time, a road vehicle propelled by a four-stroke gasoline engine

Figure 30. Benz Patent Motorwagen, 1886. In contrast to the Daimler/Maybach motor carriage (Figure 28), Benz based his construction on the front-steered three-wheeled cycles popular in Britain and France. Chain drive, tubular steel frame, and wire spoke wheels all originated in bicycle practice. The horizontally mounted flywheel is unusual and also provided the means for hand-starting the engine. Rear-mounted single-cylinder engine, 984 cc, 0.88 hp (0.65 kW).

moved under its own power. None of his predecessors could make the same claim: not Delamare-Deboutteville of France, nor Bernardi of Italy, or Marcus of Vienna, and most certainly not an American patent attorney named Selden. Therefore, Benz is the inventor of the automobile powered by an internal-combustion engine.

Further Developments at Daimler and Benz to the Turn of the Century

In contrast to earlier centuries, when mechanics and academics built devices and technical apparatus for their own amusement, for scientific purposes, or to satisfy the whims of aristo-cratic patrons, the inventions of the Industrial Age were aimed at the marketplace. Neither Daimler nor Benz was aware of the other's work, and therefore each must have had an interest in selling his motorcars or engines. It appears that the 1888 Munich "Power and Working Machinery Exhibition" (Kraft- und Arbeitsmaschinen-Ausstellung) indeed led to a sales contract, although that sale was not consummated. A few days after the contract was signed, the buyer's father declared it void because his son had been behaving irrationally of late and was committed to an asylum even before the car could be delivered.

At the 1889 Paris Exposition, both Benz and Daimler hoped to achieve a breakthrough from being a mere local to an international phenomenon. However, at the exhibition, the general public hardly noticed the Benz car or Maybach's new "Stahlradwagen" ("Steel-Wheeled Car," Figure 31). There were too many distractions and interesting cultural achievements, quite apart from the newly completed Eiffel Tower, an iron structure that soared 300 meters (984 ft) into the sky and dominated every conversation and news dispatch.

In contrast to Daimler, who preferred to see his engines installed in conventional coaches, Maybach had an eye toward commercially available quadricycles and pressed for a more convincing mechanical arrangement for future motorcars. Indeed, at the Paris Exhibition, he labeled his four-wheeler a quadricycle; the "Steel-Wheeled Car" name did not come into popular use until later. Maybach adopted the design features of contemporary pedal cars, including fork steering of both front wheels, tubular steel frame, and steel spoke wheels. Similar to the first Benz "Patent Motorcar" of 1886, this vehicle would be unthinkable without bicycle technology.

Public indifference aside, the steel-wheeled car caught the attention of French entrepreneurs. The owners of the firm Panhard & Levassor (P&L), manufacturer of woodworking machin-ery, obtained licenses from Daimler and kept the World's Fair display vehicle in Paris.

With permission of the licenser, they assigned the rights to build Daimler engines to the Peugeot tool and bicycle works. In 1890, Peugeot developed its own quadricycle (Figure 32), a faithful copy of the Maybach steel-wheeled car and, more importantly, France's first practical internal combustion road vehicle. Motivated by Peugeot's success, P&L began production of its own gasoline-engine cars following Daimler's license.

Figure 31. Daimler/Maybach Stahlradwagen, 1889. The Stahlradwagen (Steel-Wheeled Car), originally known as the quadricycle, was also based on bicycle practice. The frame, built by a company that would later be known as NSU, also served to carry cooling water. Other noteworthy features include a change gear transmission and a multi-cylinder reciprocating engine, both unprecedented in vehicle design. V2 engine, mounted upright at the rear, 565 cc, 1.5 hp (1.1 kW).

Figure 32. Peugeot quadricycle, 1890. Closely following the design of Daimler/Maybach's Stahlradwagen, Peugeot's first gasoline-powered vehicle formed the basis for the firm's subsequent models. The quadricycle was powered by a license-built Daimler engine, already hidden under a "hood." V2 Daimler engine, mounted upright at the rear, 565 cc, 1.5 hp (1.1 kW).

Even more than the Eiffel Tower and Edison's phonograph, also exhibited at the World's Fair, the ripples sent by the steel-wheeled car transformed the lives of subsequent generations, including our own. The steel-wheeled car marks the birth of the French auto industry, which triggered the first wave of motorization in Europe, with corresponding effects in North America.

For Benz, the World's Fair at first seemed less of a success than it had been for Daimler, a fact that may be traced to the three-wheeled design of his car. In 1893, presumably at the urging of his French agent Emile Roger, Benz, after reinvention of Ackermann steering (known since 1816), introduced the four-wheeled "Viktoria" (Figure 33).

Figure 33. The Benz Viktoria, 1893. The Viktoria, derived from horse-drawn carriage practice, was the first four-wheeled Benz vehicle, fitted with Ackermann steering reinvented by Benz himself. Customers could order an additional bench seat over the dashboard, in which case passengers sat facing each other (vis-à-vis). The Viktoria used a steering wheel in place of tiller steering. This photograph shows Carl and Berta Benz on an excursion. Single-cylinder engine, mounted horizontally at the rear, 1724 cc, 3 hp (2.2 kW).

The second conspicuous feature of the Viktoria is that it followed conventional carriage-making principles; that is, Benz had abandoned the bicycle-derived construction of his first three-wheeled motorcars. The situation at Maybach and Daimler was even more confusing: their motor carriage of 1887 was followed by their cycle-like steel-wheeled car of 1889, in turn followed by the carriage-like Riemenwagen ("Belt Drive Car") of 1895. There seemed to be great uncertainty about a definitive design pattern for the motorcar.

Around the same time that the Viktoria was introduced, Benz developed a small car named the "Velo," which, as indicated by its root, the velocipede, again drew on bicycle design. Despite its somewhat antiquated appearance, the Velo (Figure 34), with rear-mounted engine and belt drive, developed into the most successful vehicle of its time, thanks to a relatively favorable price of 2000 Marks. Between 1894 and 1902, Benz sold 1,200 examples. By the turn of the century, the Benz works had grown to the world's largest auto manufacturer, ahead of Daimler, Peugeot, and Panhard & Levassor.

Figure 34. Benz Velo, 1894. The Velo was a small car fitted with block brakes, belt drive, and wooden frame, and it seemed somewhat antiquated even at its introduction. However, its low price of only 2000 Marks attracted many customers domestically and abroad, especially in France. Single-cylinder engine, mounted horizontally at the rear, 1045 cc, 1.5 hp (1.1 kW).

France as the Pacemaker of Motorization

Despite German improvements such as the spray carburetor (1893, see Appendix), the first four-cylinder engine (1890), the "Phoenix" V-twin engine of 1892, the tube-core radiator (1897, see Appendix, Tubular radiator), all by Maybach, and Bosch make-and-break ignition (1897, see Appendix, Ignition systems), beginning around 1891 France took the lead in every aspect of the automobile. In Germany, public attitudes, poor roads, and limited buying power stood in opposition to mass motorization. However, the French accepted the automobile with less prejudice and saw in it a bright future. As a result, Benz and Daimler were able to export a considerable portion of their production to France.

The motorcars of Peugeot and Panhard & Levassor, initially resembling the designs of Maybach, soon developed their own unique character and exerted their own influence on automotive technology. The drivetrain layout of Panhard & Levassor, with an upright engine at the front and rear-wheel drive (Figure 35), would henceforth be the standard by which other cars were judged. This configuration put French manufacturers years ahead of Benz and Daimler. Pneumatic tires,

Figure 35. Panhard & Levassor, 1891. With a front-mounted upright engine and rear-wheel drive, Panhard & Levassor established the "standard" layout, although Maybach gave the automobile its unique appearance with his 1900/1901 Mercedes. The Panhard & Levassor displayed improved road manners due to better weight distribution, but it did not yet have a front-mounted radiator and instead used a rear-mounted water tank. V2 Daimler engine mounted upright at the front, 565 cc, 2 hp (1.5 kW).

first tried by André and Edouard Michelin on a Peugeot in 1895, allowed higher speeds with the same engine power, absorbed slight road irregularities, and permitted lighter vehicle designs. The surprising advancement of automotive technology in its early years would not have been possible without pneumatic tires. The concept found its way to production cars the following year, on the cars of de Dion-Bouton (Figure 36) and Léon Bollée (Figure 37). Michelin led the way, even before Continental and Metzeler in Germany and Dunlop in Britain, who began tire production in 1898 and 1902, respectively.

In the mid-1890s, cycling was a popular sport in both England and France, especially after Englishman John Kemp Starley built his Rover safety bicycle in 1885 (essentially our modern bicycle) and John Boyd Dunlop (1840–1921) provided pneumatic tires for it in 1888. Bicycling as a popular sport was a necessary precursor to motorization. It gave thousands of cyclists experience with fast and easily maneuverable road vehicles, awakened a desire for individual transport, and rewarded the ability to maintain, understand, repair, and reproduce a technical object.

Figure 36. The de Dion-Bouton tricycle, 1899. Beginning in 1895, engines designed by Bouton, with speeds of approximately 1500 rpm and higher specific output, may be regarded as further developments of Daimler/Maybach's medium-speed engines of approximately 700 rpm. In addition to supplying engines for custom installation, de Dion-Bouton built its own tricycles as of 1896 and quadricycles from 1898, both with pneumatic tires, with optional trailers. Single-cylinder engine mounted upright behind the rear axle, 402 cc, 3.5 hp (2.6 kW).

Figure 37. Léon Bollée Voiturette, 1896. Concurrent with the de Dion-Bouton, the tandem three-wheeler was the first series-produced road vehicle with pneumatic tires (by Michelin) and was propelled by a powerful, but not especially reliable, engine. Hot-tube ignition and spray carburetor followed Daimler/Maybach practice. Single-cylinder engine mounted horizontally alongside the rear wheel, 922 cc, 2.5 hp (1.8 kW).

France's excellent roads, which dated to the rule of Napoleon, lent themselves not only to bicycle races but also to automotive contests. For manufacturers, victory represented the best possible advertisement; for the engineers and mechanics, vindication; and for the spectators, thrills not unlike those of the chariot races of ancient Rome.

In 1894, using bicycle races as a pattern, journalist Pierre Giffard of *Le Petit Journal* organized the first automobile road race, stretching from Paris to Rouen (a distance of 126 km [78.3 miles]). The main criteria were reliability and utility, rather than outright speed. Of 102 entries, 21 cars survived qualification trials to take part in the race. A de Dion-Bouton steam tractor towing a carriage finished first but could not fulfill the other criteria; therefore, overall victory was awarded jointly to the gasoline-engined motorcars of Peugeot and Panhard & Levassor, both powered by Daimler engines. These achieved average speeds of approximately 18 km/h (11 mph), approaching those of bicycle racers.

A pure speed contest for automobiles took place a year later, the first in history. The 1200-km (746-mile) marathon run from Paris to Bordeaux and back saw a field of fifteen gasoline-powered cars, six steamers, and one electric car, as well as two gasoline motorcycles. Panhards and Peugeots took the top placings, again powered by license-built Daimler engines, followed by a Roger-Benz. The average speed of victor Emile Levassor was a respectable 24 km/h (15 mph). In proving the superiority of the gasoline engine over all other power sources, the race represented a triumph for Daimler/Maybach and Benz.

This prompted Count Albert de Dion to give up construction of steam cars and concentrate instead on gasoline-powered cars and gasoline engines. His chief engineer Georges Bouton built a high-speed, high-specific-output, lightweight single-cylinder engine. In 1899, to increase his technological advantage over the license-built Daimler engines, de Dion set up a metallurgical laboratory, probably the first research and development center in the auto industry. With an estimated production of 26,000 engines by 1902 and thousands of tricycles and quadricycles (Figure 36), de Dion-Bouton by the turn of the century had grown to be the world's largest producer of aftermarket engines and complete vehicles. Thus, de Dion-Bouton must be considered the company most responsible for the motorization of Europe.

The race of 1895 was the first of a long line of road races starting and ending in Paris. Their effect on the general public must have been considerable. In 1890, only four manufacturers of gasoline automobiles existed worldwide: Benz, Daimler, Peugeot, and Panhard & Levassor. By 1899, their number had swelled to 75 in France alone. Races of that era were usually won by the products of Panhard & Levassor or Emile Mors. The cars of Mors were considered to be on the cutting edge of technology. He developed his own ignition systems, dry-sump lubrication systems (see Appendix), and shock absorbers. Class victories were won by Louis Renault, whose first car of 1898 boasted a driveshaft instead of chains and their attendant oil splatter, and a gear change transmission with a direct top gear (see Appendix) instead of a belt drive.

The situation in France, and the bicycle crisis of 1898, caused by overproduction and cheap imports from the United States, prompted several German entrepreneurs to take up automobile production before the turn of the century. They drew heavily on French technology, and occasionally on Benz and Daimler. Adler took Renault as his inspiration. Cudell of Aachen obtained licenses from de Dion-Bouton. Dürkopp in Bielefeld was a licensee of Panhard & Levassor. Fahrzeugfabrik Eisenach (the Eisenach Vehicle Works) and Falke of Mönchen-Gladbach were licensed by Decauville. Lutzmann and Süddeutsche Automobilfabrik Gaggenau (Gaggenau South German Automobile Works) showed Benz influences, while the products of Motorfahrzeug- und Motorenfabrik Berlin-Marienfelde (the Berlin-Marienfelde Motor Vehicle and Engine Works) were illegal copies of Daimler. Opel started by building vehicles based on a Lutzmann patent (Figure 38); beginning in 1902, the firm assembled French Darracqs.

Figure 38. Opel Lutzmann, 1899. On January 21, 1899, sewing machine and bicycle manufacturer Adam Opel obtained the manufacturing rights to the Lutzmann motorcar, designed by Friedrich Lutzmann along Benz Viktoria lines. The two-seat model was priced at 3450 Marks, the four-seater at 3800 Marks. Optional pneumatic tires added 250 and 350 Marks to each model, respectively. Single-cylinder engine mounted horizontally at the rear, 1542 cc, 3.5 hp (2.6 kW).

Europe and North America Embrace the Idea of the Automobile

By the turn of the century, worldwide interest in the automobile was enjoying phenomenal growth. The year 1899 saw the founding of the Österreichische Daimler-Motoren KG (Austrian Daimler Engine Co.), later known as Austro-Daimler. The Nesselsdorfer Wagenbau-Fabriks-Gesellschaft (Nesselsdorfer Wagonbuilding Works Inc.), later Tatra, built its Präsident model along Benz lines (Figure 39). In Vienna, Jacob Lohner, builder of carriages for the Austrian court, offered gasoline, electric (Figure 40), and hybrid (gasoline-electric) power, so-called "Mixte cars." Names appeared which would one day influence central European automotive technology: Edmund Rumpler (1872–1940) and Hans Ledwinka (1878–1967) contributed to the making of the Präsident; Ferdinand Porsche (1875–1951) improved the electric motor of the 1898 Lohner electromobile; Joseph Vollmer (1871–1955) worked for Gaggenau, Kühlstein, and NAG. He designed the DURCH tractor trailer of 1903 (Figure 67) and the first German tank of 1917.

Siegfried Marcus (1831–1898) remained an outsider. For a long time, part of the automotive world regarded him as the inventor of the automobile, as indicated by the year 1877 claimed on

Figure 39. Nesselsdorf Präsident, 1898. The first motorcar produced by what would later be known as the Tatra Works, powered by a Benz Kontra engine. The handlebar steering arrangement also served as a transmission shift lever. Note front bodywork with bumper. Two-cylinder boxer (horizontally opposed) engine (Benz) mounted at the rear, 2714 cc, 5 hp (3.7 kW).

a display panel at the 1898 Vienna Commercial Exhibition. If true, he would have been about ten years ahead of Benz and Daimler. Only in 1968 was it determined that the vehicle known as the second Marcus car (Figure 41) originated in 1888. His first car of 1870 was a non-steering handcart on which Marcus had installed an atmospheric two-stroke gasoline engine. In 1882, Marcus designed a small and effective carburetor, and in 1883 magneto ignition in a form similar to Bosch's design fourteen years later. Claims of precedence aside, with his 1888 car, Marcus at least deserves recognition for having the first car with a magneto-ignition four-stroke engine.

Other nations such as Belgium, Switzerland, Italy, Sweden, and the Netherlands began their own automotive activities with designs based on Benz, Daimler, Peugeot, or Panhard & Levassor. Only in Britain was any sort of activity involving mechanically driven vehicles prohibited by law. The Red Flag Act had been extended to include gasoline-powered vehicles, and it stifled all initiative.

41

Figure 40. Lohner electric fire engine, circa 1905. In addition to passenger cars, Lohner also offered commercial vehicles with battery-electric drives. The batteries could be mounted above the front axle, under a cover resembling a conventional engine hood, or in panniers below the frame. In either case, propulsion was by means of hub motors, designed by Ferdinand Porsche, in the front wheels. Available bodies included fire engine, crew transport, or custom designs. Power output 12–15 kW.

Figure 41. Marcus' car, 1888. Goose-necked wooden frame members connected by transverse braces carried turntable steering at the front, and amidship-mounted engine. Power was transmitted to the rear axle through a steel cone clutch and circular cross-section belts. The engine was equipped with low-tension magneto ignition (see Appendix, Ignition systems) and a brush carburetor patented by Marcus. Horizontal mid-mounted single-cylinder engine, 1570 cc, 0.75 hp (0.55 kW).

When the Red Flag Act was finally repealed in 1896, England, which beginning in 1802 had led the world in steam-powered road vehicles, was ten years behind the times. The first automobiles on British roads had to be imported from France and Germany. The motorcars offered by Daimler Motors Syndicate Ltd. in Coventry were actually imported from Daimler-Cannstatt. The English Arnold, Marshall, and Star were Benz copies. Shortly thereafter, as the predecessors of today's London Transport switched from horse-drawn to motorized transport, omnibuses and bus chassis had to be imported. Frederick Lanchester (1868–1946) was probably the only constructor who worked without Continental examples, as he began to build cars with opposed engines, driveshafts, and worm gear drives (see Appendix, Worm gear drive).

At the end of the nineteenth century, while European car makers could count on a steady albeit modest turnover, American would-be car makers found themselves in the experimental stage. Preconditions for motorization were less than favorable. Legislation and court verdicts generally stood in opposition to mechanically driven vehicles. Unlike Europe, cross-country roads were nonexistent in America. Beginning around 1850, the North American continent was opened by the steam railroad, which also consumed most of the available investment capital.

Another steam-powered device was the road vehicle developed by Oliver Evans, which he presumably sought to put into service as a means of transport in and around Philadelphia. When he was unable to secure funding, Evans rebuilt his "Orukter Amphibolos" (Figure 42) as a steam shovel and anchored it in the Schuylkill River, where it became known as the "Mud Machine."

Countless steam-powered wagons, motorcycles, and buggies followed (see Appendix), which, despite their complex operational requirements, were superior to the electric vehicles of the time. Prior to the invention of the lead-acid battery around 1860 and rapid charging, which enabled "proper" electric vehicles, electrics were dependent on cables, electrified tracks, or galvanic elements for their power supply.

After Belgian Camille Jenatzy (1868–1913) became the first to exceed 100 km/h (62 mph), driving a streamlined electric car (Figure 58) near Paris in 1899, many Americans felt that only electric drive (see Appendix) had any chance of success. They accepted steam power; however, internal combustion engines, with their problematical auxiliary devices, were considered a technological dead end.

While Europe had already settled on the gasoline engine in the 1880s, in the United States various systems were competing for supremacy well beyond the turn of the century. Beside compressed air, acetylene, and even spring power, the manufacturers of steam, electric, and gasoline-powered cars struggled to capture the burgeoning market for themselves.

Inspired by an imported de Dion-Bouton steam-powered tricycle, the Stanley twins, until that time makers of violins and photographic plates, built their 1897 steam buggy, the prototype for an entire line of competing models. A year later, the Stanleys sold several patents to two

The Mechanic.

JULY, 1834.

[For the Mechanic.]
STEAM-CARRIAGES.

Figure 42. Orukter Amphibolos, 1805. Inspection of drawings published in 1834 raise more questions than answers. If the lines connecting flywheel and the front and rear wheels are intended to represent belts, cables, or chains, the Orukter could travel only in a straight line. If floating in water, the paddle wheel is mounted below the water-line—too low to be functional. The shovel arrangement is missing. Later, a model was built based on Oliver Evans' written description, differing from the drawing in many details. Today, it is preserved in the Smithsonian Institution in Washington, D.C.

entrepreneurs who founded Locomobile and the Mobile Company of America (Figure 43). Both firms gave up on steam after several years; however, Stanley, the leader of the steamer contingent, soldiered on until 1925. Moreover, in 1906, racer Fred H. Marriott drove his streamlined Stanley "Wobblebug" to a new speed record of 195.652 km/h (121.573 mph) which stood for three years.

From the mob of about 80 American steam car builders, the White Company, erstwhile makers of sewing machines, developed as Stanley's most persistent competitor. Thanks to compound engines, condensers, and flash boilers inspired by Serpollet's ideas, the firm improved engine output and acceleration and decreased water consumption. President Theodore Roosevelt honored their technical superiority, being chauffeured in a White as of 1907. The honor of being the only steam car manufacturer to be chosen by the President of the United States did not cloud the White brothers' view of the bigger picture. In 1910, they ceased steam car production in favor of gasoline power and concentrated on manufacturing commercial vehicles.

Figure 43. Mobile, 1900. The Mobile Company of America, located in Tarrytown, New York, built light steam cars with petroleum-fueled boilers, two-cylinder engines, steel tube frames, and wire spoke wheels. Time to raise steam was approximately 15 minutes; the water supply was sufficient for a range of approximately 50 km (31 miles). Approximately 6,000 examples were built during its production life, from 1899 to 1903. Two-cylinder mid-mounted upright engine, 6 hp (4.4 kW).

Electric cars found greater favor among women motorists, doctors, and taxi companies. In the words of the 1907 *Automobiltechnischer Kalender*, "Its cleanliness, its elegance and its silent progress have made the electromobile an important part of modern city traffic." Indirectly, these words underscore the major handicap of wireless electric vehicles: rapid discharge of the batteries rules out any sort of cross-country travel. The fervent hope of the time—that manufacturers would soon introduce lighter, more durable, and higher-performance batteries—has not been fulfilled to this day. As early as 1900, well before their gasoline counterparts, American electric cars had electric illumination (Waverly), enclosed bodywork (Detroit Electric, Studebaker), driveshafts instead of chains (Columbia, Baker), and elimination of the tedious hand-cranking process that made gasoline cars so unacceptable for the "fairer sex." In general, electrics were equipped more luxuriously and were more pleasant to drive than representatives of the steam and gasoline schools.

In the United States, gasoline engines arrived late on the scene. In 1891, John William Lambert and Henry Nadig undertook epic test drives with their automobiles, five years after the first American electric cars and the arrival of gasoline motorcars on the European scene. None of the subsequent pioneers drew on Daimler engine experience, although these were sold in New York from 1891 by a firm founded by piano builder William Steinway. George Bailey Brayton could also have delivered finished engines; the Brayton Ready, with its air injection, was on the market in 1876 and is considered the first mass-produced gasoline engine. Selden's vehicle was powered by a Brayton engine.

Figure 44. Selden buggy, 1906. At a time when the Curved Dash Oldsmobile (Figure 49) was already in mass production, a court verdict forced Selden to actually build (in fact, to have built) an example of the "forecarriage" described in his 1879 patent application, to prove that his design was even capable of moving under its own power. The front-drive buggy indeed ran, even if only for a short distance. Today, it is on display in the Henry Ford Museum in Dearborn, Michigan. The year 1877 on the vehicle refers to Selden's first design considerations and sketches. Three-cylinder Brayton-cycle two-stroke engine, mounted horizontally over the front axle.

In 1879, American attorney George Baldwin Selden filed a patent for an "auto buggy" (Figure 44). By means of well-timed amendments and dramatic pauses, Selden managed to delay granting of the patent until 1895. Using such tactics, Selden not only claimed to be the inventor of the automobile, but also employed the Association of Licensed Automobile Manufacturers—the ALAM— to extort licensing fees from approximately 30 auto manufacturers and importers, without ever having built the buggy in question. This oversight was corrected by court order in 1906. After years of legal wrangling between Selden and a group of auto makers led by Henry Ford, the New York Supreme Court voided the Selden patent in 1911.

With his legal barrage, Selden delayed American development of the automobile by years. According to *The Horseless Age*, 80 mechanically driven road vehicles were operating in the United States in 1895. Of these, 50 percent were gasoline powered, 17 percent were electric, and 13 percent were steam powered. Five years later, the proportion of gasoline power in the 4,192 vehicles built to date had dropped to 20 percent. Electric and steam power each held 40 percent of the market. In the United States, the gasoline engine did not gain acceptance until after 1900, in part influenced by the success of French automobiles and the 1900/1901 Mercedes.

The catalyst for American motorization was an event organized in 1895 by the *Chicago Times-Herald*, less a race than an assessment of the state of the automotive art, a test, and a stimulant to further development. Measuring power output of the vehicles present by means of a dynamometer was intended to increase understanding of mechanical design relationships. The subsequent race was contested by two electrics, three Benz that had been modified by their owners, and J. Frank Duryea with a car of his own design (Figure 45). Duryea won, followed home by Charles Brady King aboard a Mueller-Benz. None of the other entrants finished. The *Chicago Times-Herald* contest encouraged, among others, Henry Ford, Alexander Winton, and Ransom E. Olds to build their own motorcars.

Figure 45. Duryea, 1895. J. Frank Duryea participated in the first American auto-mobile race, the Chicago Times-Herald *contest of 1895, driving a motorcar of his own design. He won the 88 km (54.7 mile) race under winter conditions, with an average speed of approximately 12 km/h (7.5 mph). Single-cylinder engine, mounted horizontally at the rear, 4 hp (2.9 kW).*

The Automobile Industry at the Turn of the Century, I

Prior to 1900, the automobile was neither an economic nor a social factor, at least not in Germany. While King Edward VII introduced the automobile to the British court in 1900, Kaiser Wilhelm II exhibited indifference. German aristocracy, which kept an eye on the activities of the royal household but otherwise spent more time on their country estates than in Berlin, cultivated horse breeding and maintained carriages. Their attitudes and activities in turn influenced an upper class rather lacking in self-confidence. However, for the ordinary citizen, an automobile or even a motorized three- or four-wheeled cycle was virtually unaffordable. In 1898, a 2.75-hp Benz Velo, a rather modest conveyance, cost 2200 Marks. A tricycle of the de Dion-Bouton pattern was priced around 1500 Marks. An ordinary worker earned approximately 60 Marks per month with a 62-hour work week. Most of the motor vehicles produced in German factories were exported, chiefly to France.

By contrast, a bicycle cost 150 to 200 Marks ($35–$47, in 1900) bringing it within reach of the ordinary man. Bicycle production numbers were correspondingly high. During the boom between October 1896 and September 1897, the United States built 900,000, England built 500,000, Germany built 350,000, and France built approximately 100,000 bicycles. By comparison, until 1899, automobile production in these countries remained well below 1,000 vehicles per year. Broken down by company, production figures were as follows: Peugeot built 8,000 bicycles against 29 motorcars (1892), Dürkopp made 50,000 bicycles and a couple of motorcars in the 1896–1897 season, Opel produced 16,000 bicycles in 1898 but only 11 cars in the following year, Stoewer assembled 13,500 bicycles in 1896–1897, while both Adler and Reichstein/Brennabor each produced approximately 30,000 bicycles in 1898. (All numbers are according to company claims.)

Socially, at least in Germany, the automobile was overshadowed by the bicycle, and as a matter of transportation policy took a back seat to the railways. By 1899, German state and private railways had expanded their track network to approximately 48,000 km (29,827 miles), putting them in second place behind the United States.

In terms of railway rolling stock for passenger service, Germany again placed second, behind Great Britain. With their imposing facades and gracefully arched steel and glass halls, railway stations dominated the centers of all major cities. The headquarters of automobile clubs, which sprang up after 1900, remained ordinary buildings among all the other ordinary buildings in the cityscape.

When the automobile, "this queer-looking thing, a kind of caricature, first made its appearance, nobody knew whether to laugh at it as a good joke or to become indignant at the intrusion...In all its picturesque ugliness, it is a boon and a blessing...We not only give it our respect but our admiration, for with its big rubber wheels it gets over the ground in a velvety sort of way and reaches its destination without becoming tired...All hail to the automobile, and may some gifted genius soon arrive who will whip it into shape and make it presentable! All things are possible, even a good-looking horseless carriage." (*Scientific American*, January 21, 1899, p. 39)

The Mercedes Era

America's longing for a good-looking automobile was fulfilled sooner than expected. In 1899, to counter seemingly insurmountable French competition, Emil Jellinek, the Daimler agent for southern France, prompted construction of a racer that was not only more powerful but also safer than the previous examples from Daimler and its competitors. Although Gottlieb Daimler opposed building special race cars, Maybach began design of such a car (Figure 46) even while Daimler was still alive. (Daimler passed away in March 1900). Its performance capabilities and visual appearance set automobile design standards for the next several decades.

Figure 46. Mercedes 35 HP, 1901. Wilhelm Maybach designed the first Daimler to carry the Mercedes marque in slightly less than ten months. During the March 1901 race week in Nice, France, it simply ran away from all other competitors. Shortly thereafter, the two-seat racer reached the marketplace as a four-seater, otherwise virtually unchanged (as shown here). This vehicle marks the beginning of the modern automobile. Four-cylinder engine, mounted upright at the front, 5918 cc, 35 hp (26 kW).

Maybach, whom the French admiringly called "le Roi des Constructeurs," began with a lowered chassis and extended wheelbase, in which he installed a high-performance engine (Table 1) incorporating numerous technical advances: an aluminum alloy crankcase, dual camshafts, crossflow head, Bosch make-and-break ignition, and a newly developed honeycomb radiator (see Appendix) whose increased surface area was coupled with a reduction in water volume, radiator dimensions, and weight.

TABLE 1
COMPARISON OF DMG FOUR-CYLINDER ENGINES

	Daimler 1898	Mercedes 1900
Output, hp @ rpm	23/900	35/1000
Displacement, cc	5295	5918
Engine weight, kg	322	238
Specific output, hp/liter	4.2	5.9
Specific weight, kg/hp	14	6.8

NOTE: Specific output is the power output per liter of displacement. It serves to compare engines of different sizes and is expressed in horsepower per liter or kilowatts per liter. High numbers are more desirable. Specific weight is the power output per unit of (dry) engine weight and is expressed in kg/hp or kg/kW. Low numbers are more desirable.

Maybach's design, which for marketing reasons Jellinek named for his eldest daughter Mercédès rather than the customary Daimler label, handily beat all opposition during the 1901 Nice race week. In his retrospective of the automotive events of that year, Paul Meyan, general secretary of the Automobile Club de France, wrote "Nous sommes entrés dans l'ère Mercédès" ("We have entered the era of Mercedes").*

French and Italian competitors rushed to copy Maybach's design details. They did this so successfully that Mercedes was unable to garner any further victories. It was not until 1903 that Camille Jenatzy, driving a 60-hp Mercedes (Figure 47) was able to beat teams representing Great Britain, France, and the United States in the Gordon Bennett Cup. The Daimler Motoren Gesellschaft had achieved its breakthrough into the international motorsports scene. It was precisely this "traveling trophy," sponsored by American newspaper publisher James Gordon Bennett, Jr. from 1899 onward "to encourage the sport of motor racing" (1904 Gordon Bennett guidebook) that opened the American export market to DMG.

Mercedes' influence on overall development of the automobile also manifested itself in unexpected ways. Firms throughout the world (Figure 48) copied the Mercedes form and function—and often learned the hard way that clutches, chassis, springs, and many other components could not cope with the side effects of more powerful engines. One example is frame members; tubular frames, adopted from bicycle technology, proved too weak and had to be replaced by U-channel frames following Mercedes practice. Firms that had responded to the bicycle crisis of 1898–1899 by shifting production to that related new product, the automobile, now faced the realization that they would have to invest in expensive stamping and shearing dies, buy their frames from suppliers, or give up auto making entirely. Mercedes not only unleashed the first great wave of capital investment in the then young auto industry; it also sealed the fate of manufacturers that had inadequate capitalization.

At Benz, the world's largest auto maker until 1900, sales figures plummeted. In 1901, Peugeot overtook Benz, followed by Daimler in 1903 (Table 2), to say nothing of American firms. Carl

---

* F. Schildberger: *Gottlieb Daimler, Wilhelm Maybach und Karl Benz*, Stuttgart, 1968, p. 50.

Figure 47. Mercedes 60 HP, 1903. Camille Jenatzy won the fourth Gordon Bennett Race, held in 1903 in Ireland, at the wheel of a Mercedes with an average speed of 89.2 km/h (55.4 mph), ahead of two Panhards and a Mors. For better recognition, the cars of individual nations appeared in various colors which, in some cases, developed into national racing colors. British cars were painted green, French cars light blue, American cars red, and German cars white. Four-cylinder engine, 9235 cc, 60 hp (44 kW).

Figure 48. Royce 10 HP, 1904. With the exception of Lanchester, before the turn of the century, British motorcar builders followed the examples of Benz, Daimler, and the French manufacturers. So too Henry Royce, who configured his first car following Mercedes practice. Its front end design, with minor modifications, was retained by succeeding generations of Rolls-Royce for decades. Two-cylinder engine, mounted upright at the front, 1800 cc, 10 hp (7.4 kW)

Benz tried in vain to catch up to his competition. The board of directors of Benz & Cie entrusted the design of new engines to a French engineer, Marius Barbarou, a situation that Carl Benz found intolerable. He left the gas engine works in 1903; two years later, he founded C. Benz & Söhne in Ladenburg, near Mannheim. The goal of the firm was manufacture of engines and motorcars, but the company was never of any great significance.

TABLE 2
PRODUCTION FIGURES, 1894–1903

	Benz	Daimler*	Peugeot	Opel	Fiat
1894	67	1	40	–	–
1895	135	8	72	–	–
1896	181	24	92	–	–
1897	256	26	54	–	–
1898	434	57	156	–	–
1899	572	108	323	11	8
1900	603	96	500	24	24
1901	385	144	456	30	73
1902	226	197	637	64	107
1903	173	232	773	178	134

* Depending on source, 322, 329, or 345 cars are given for the period 1887–1900.

The Great Awakening

While the Europeans saw the automobile as a technological challenge, Americans regarded it above all as a means of making money. The prolifically polygamous sewing machine tycoon Isaac Singer expressed the prevailing philosophy in this way: "I don't care a damn for the invention. The dimes are what I'm after." What mattered was units sold per year, an American phenomenon that led from mass production to the assembly line.

In the nineteenth century, several American factories had begun mass production of complex articles including guns, clocks, sewing machines, and bicycles; automobiles followed beginning around 1900. In that year, the Columbia & Electric Vehicle Co. produced 1,500 electric cars, while Locomobile built 750 steamers. The first mass-produced car was the Curved Dash Oldsmobile (Figure 49), built by Ransom Eli Olds. Between 1900 and 1904, Olds made 11,275 examples. The Curved Dash also marks the ascendancy of the gasoline engine over other power sources.

Despite much higher volume compared to Europe, it soon became obvious that future domestic demand could not be met using conventional production methods. Comprising approximately 5,000 individual parts, the automobile was one of the most complex industrial products of its time. In contrast to other technically sophisticated products such as machine tools and locomotives, automobiles had to be built in greater numbers, in less time, and at lower prices.

There were definite limits to growth. Factories could not be expanded indefinitely. Until World War I, the United States was a debtor nation, and investment capital was limited. Shortage of

Gazelle-Motorwagen
Hocheleganter Zweisitzer
mit kleinem Vordersitze für Kinder.

Preis 3000 Mark.

2 Vorwärtsgänge und 1 Rücklauf mit nur **einem** Steuerhebel einzuschalten.
1 Fussbremse auf das Vorgelege und 1 Handbremse direkt auf die Hinter-
radachse wirkend. Der Vergaser hat Rückschlagventil, ist explosionssicher
und sehr ökonomisch. Benzin-Motor ca. 6 PS, eincylindrig mit elektr. Zündung
durch Akkumulatoren. Eine Benzinfüllung (15 Liter) reicht bis zu 150 Kilometer.
Einstellbar auf 5 - 35 Kilometer per Stunde, ausgezeichneter Bergsteiger.
Denkbar einfachste Handhabung.
Stossfreier, geräuschloser Gang. Solide Konstruktion. Beste Pneumatiks.

Polyphon-Musikwerke Aktien-Gesellschaft,
Wahren bei Leipzig.
(Abteilung Automobilbau.)

Figure 49. Polymobil Gazelle (license-built Oldsmobile), 1904. Through its New York offices, the Polyphon-Musikwerke AG (Polyphon Music Works Inc.) obtained the German license to build the Curved Dash Oldsmobile at its plant in Wahren near Leipzig. Technically simple yet affordable and rugged, the car was sold in Germany as the Polymobil Gazelle. The bodywork, slightly modified from the original, retained the curved splash guard (dash-board), which gave the model its American name. Single-cylinder engine, mounted horizontally at the rear, 1543 cc, approximately 6 hp (4.4 kW).

labor resulted in mechanization of nearly all branches of industry. Although the degree of mecha-nization of the American automobile industry was higher than that of European makers, it was still too low to cover domestic demand (Table 3).

TABLE 3
PRODUCTIVITY PER WORKER, 1907

	Employees	Production (Units)	Productivity (Units per Man-Year)
Packard	4,640	1,403	0.3
Cadillac	3,500	2,884	0.8
Buick	4,000	4,641	1.2
Ford*	2,595	14,887	5.7
For comparison:			
Daimler 1915			1.5
Ford (USA) 1977			12.0

NOTE: Figures are approximate.
* In 1907, Ford was largely an assembly plant, needing fewer workers.

More than any other manufacturer, Henry Ford faced the problem of increasing his daily output. Demand for his Model T (Figure 50), which was introduced in 1908, forced him to build cars on a moving assembly line, a concept that had already been applied in Cincinnati stockyards as of 1869, the canned goods industry, and a Pittsburgh foundry. After the principle of interchangeable parts was promulgated by Cadillac, Ford streamlined work processes following the concepts of Frederick Winslow Taylor. The first Ford assembly line was installed in December 1913.

Before the introduction of the assembly line, completion of a chassis—complete with engine, axles, and springs—took 12.5 hours; thereafter, it required only 2.6 hours (December 1913). In his 1923 autobiography, Henry Ford wrote, "In the early part of 1914 we elevated the assembly line. We had adopted the policy of man-high work...The waist-high arrangement and a further subdivision of work so that each man had fewer movements cut down the labour time per chassis to one hour thirty-three minutes." (Ford, Henry, *My Life and Work*, Doubleday, Page & Company, New York, 1922) Where 1913 production had been 208,667 examples of the Model T, in 1914 output rose to 308,162 units. At its peak in 1923, Ford churned out 1,817,891 Model T's.

Figure 50. Ford Model T, 1922. Henry Ford's frequently cited quote that the car was available in "any color so long as it's black" actually applied only to Ford Model T's built between 1914 and 1926. Customers accepted this edict because the Model T price dropped from year to year. From the beginning of production in 1908 to 1914, and again in its last year 1927, other colors were available. Four-cylinder engine, 2894 cc, 20 hp (15 kW).

With the "Tin Lizzy," American industrialization enjoyed newfound respectability. As an entrepreneur, Ford distanced himself from the uncouth, exploiting style of the early years. He had discovered that his workers were consumers, too. He founded his claim to have democratized the automobile on three measures. First, he paid salaries that were 10 to 15 percent higher than the going rate. Second, in 1914, he introduced a guaranteed minimum wage of $5 per day. Third, he steadily reduced the price of the Model T, down to $290 in 1924.

Ford's form of capitalism integrated the assembly line worker, heretofore largely living in poverty, into a society of mass consumption, later of prosperity, indeed of excess. The workers themselves had varying attitudes toward "Fordism." They applauded the 1926 introduction of six days' pay for five days' work, and the increase of the minimum wage to $7 per day in 1929. Some assembly line workers saw the steady, repetitive nature of their tasks as cutting the soul out of their personalities; others allowed the monotony to create a spiritual distance between themselves and their work.

Both American and European manufacturers copied the assembly line system. Yet, as long as Ford's overwhelming dominance remained unassailable—the firm owned 50 percent of the American market in the 1920s—competitors had to resort to other methods. In 1921, Hudson, under the Essex nameplate, introduced a sedan built of economically produced, straight (wooden) panels and doors, which ultimately displaced the open touring car and heralded the arrival of weatherproof travel. General Motors originated the use of synthetic paints in a variety of colors and frequent model changes. In his autobiography, Henry Ford scoffed at this policy: "We have been told that this is good business, that it is clever business, that the object of business ought to be to get people to buy frequently and that it is bad business to try to make anything that will last forever..." (Ford, Henry, *My Life and Work*, Doubleday, Page & Company, New York, 1922, pp. 148–149.)

Ford claimed that his "principle of business is precisely to the contrary...We never make an improvement that renders any previous model obsolete." (Ford, Henry, *My Life and Work*, Doubleday, Page & Company, New York, 1922.) This policy ended in debacle. Technically and from a design standpoint, competitors eventually left the antiquated Model T in their dust. The Tin Lizzy became unsellable. When Model T production finally halted in May 1927, after 19 years, Ford tallied 15,007,033 examples, a production record that remained until it was broken by Volkswagen in 1972.

Because a successor for the Model T was not yet ready for production, Ford had to close his plant for half a year in 1927, which had lasting consequences. With few exceptions, General Motors' Chevrolet division would surpass Ford in the annual production numbers race.

Mass production and assembly line methods made possible by simple design, manufacturing rationalization, and profit maximization were and remain the dominant features of the American auto industry, much more than technical experimentation and individualism. Exceptions only prove the rule.

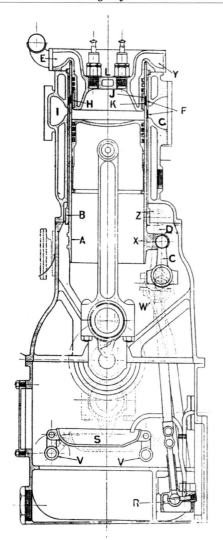

Figure 51. Daimler-(Coventry) sleeve valve engine (Knight license), circa 1911. The piston slides within an inner sleeve A, which in turn slides in an outer sleeve B. The sleeves are driven by a shaft W and connecting rod C. When ports H overlap intake manifold I, fresh mixture flows into the cylinder. When ports K match the exhaust passage G, spent gases leave the cylinder. Problems included cooling and lubrication of the sleeve A in the area of the cylinder head L, as well as the lower extremity of the sleeve.

In Syracuse, New York, Franklin built cars for the upper middle class. These departed from standard American design practice in their use of air-cooled six-cylinder engines, wooden frames, and double elliptical springs; the cars were aimed at individualistic drivers. Charles Yale Knight developed an engine employing sleeve valves instead of conventional poppet valves. Despite its low mechanical efficiency, the Knight engine (Figure 51) was built for approximately 30 years by European firms (Minerva, Mercedes, and others) and American firms including Willys-Knight, Falcon-Knight, and Stearns-Knight.

A unique development in American automotive design is represented by the so-called "high wheelers," resembling horse-drawn buggies (Figure 52). Despite primitive technology, block or friction roller brakes, cable drive, and tiller steering, they demonstrated adequate off-road ability to meet the demands of their usually rural and agrarian buyers.

Beyond the city limits, roads were often impassable. The first major east-west thoroughfare, the 5360-km (3,330-mile) long Lincoln Highway from Boston to Sacramento, was not dedicated until 1913 and even ten years later was largely unpaved. However, sportsmen were not dependent on highways; in 1903, a 20-hp Winton covered the distance between San Francisco and New York in 63 days. In 1904, a 10-hp Franklin took only 39 days, and the United States had discovered a new game. At one time or another, almost every American auto manufacturer organized a transcontinental run and milked it for all of the publicity it could.

With the growing demand for cars, manufacturers also developed a growing appetite for capital. Soon, there were joint stock companies, corporate failures, and mergers. In 1903, Albert A. Pope forged a trust whose member

Figure 52. McIntyre High Wheeler, 1909. In addition to International Harvester and Holsman (America's two largest buggy makers), smaller firms such as McIntyre of Auburn, Indiana, also offered "high wheelers." Their characteristics usually included amidship-mounted, horizontally opposed, air-cooled underfloor engines, as well as solid rubber or even iron wheels.

57

firms built bicycles, motorcycles, and gasoline and electric cars. Four years later, his concern collapsed. By 1918, John North Willys and several of the auto firms he controlled moved up to second place behind Ford. William Crapo Durant (1862–1947) bought companies by the dozen. In 1908, he founded General Motors, and in 1911, Chevrolet. After quarreling with General Motors board members, Durant founded a competing firm, Durant Motors, in 1921, which ultimately, with the same fate as so many others, went under during the Great Depression. On the other hand, General Motors developed into the largest auto company in the world.

By World War I, the "conventional" configuration of the automobile had become standard: front engine, rear-wheel drive, beam and live axles (see Figure 74). Further development improved everyday utility. Pressure lubrication (see Appendix) replaced splash oiling. High-tension ignition (see Appendix, Ignition systems) replaced make-and-break ignition or oscillating magnetos. Hand cranks gave way to electric starters. The tall, swaying motorcar of pioneering days had evolved into an automobile, with a longer wheelbase and passable handling characteristics.

Many auto manufacturers, at least in Europe, supplied only raw chassis. On these, coachbuilders erected bodywork under contract to the factory or for individual customers. Bodywork consisted of a wooden framework with wood and later metal sheathing; luxury cars employed aluminum panels (see Figure 212). In 1913, the Daimler Motoren Gesellschaft, one of the few European manufacturers with its own body plant, charged 2200 Marks for an open four-seater (a so-called "tourer"). Limousines, in which drivers still sat in the open, cost twice as much and radiated all of the charm of the Belle Epoque, particularly if they were bodied by French *carrossiers* (Figure 53). Open cars dominated the scene, as they had among horse-drawn coaches. Beginning around 1910, these types were joined by sports cars from ALFA, Austro-Daimler, Stutz, and other firms.

For luxury car makers such as Packard, Delaunay-Belleville, Rolls-Royce, Hispano-Suiza, Isotta-Fraschini, and Métallurgique (Figure 54), quality materials and impeccable workmanship were prime considerations. The lowest level of motorization on three or four wheels was represented by "cycle cars." These consisted mainly of bicycle and motorcycle parts. Their technology, with friction or belt drive (see Appendix) and steering by means of cables or chains, was often nothing short of primitive. The cycle car wave of 1910 to 1916 spawned approximately 100 firms each in England, France, and the United States, and approximately 10 in Germany. The best known were the Cyklonette (Figure 55) and the Phänomobil.

Immediately above the cycle cars on the European scene was the densely populated class to 18 hp, including models offered by Adler, Wanderer, Mathis, Hansa (Figure 56), FN, Swift, and Singer. These may be regarded as predecessors of the small cars that were so successful in the 1920s—the Austin Seven (see Figure 78), Citroën 5 CV, and Fiat 509.

For years, the somewhat more powerful touring cars, which formed the next higher class, were the center of public attention. At the instigation of German-English painter Sir Hubert von Herkomer (1849–1914), the Bavarian and German Automobile Clubs organized touring car trials. The three Herkomer Tours of 1905–1907 transitioned into the Prince Henry Trials

Figure 53. Delaunay-Belleville, 1912. Before World War I, this firm, based in St. Denis on the Seine, was one of the finest auto makers in the world. Customers included Tsar Nicholas II of Russia and Denmark's Crown Prince Valdemar. Coachwork was correspondingly graceful and exclusive. The illustration shows a coupé de ville (literally, a town car) bodied by Paul Nee of Paris-Levallois. A typical French touch is breaking up the surfaces at the sides and rear with wicker basketweave in a contrasting color. Four-cylinder engine, 2949 cc, 25 hp (18 kW).

Figure 54. Métallurgique, 1907. Based in Marchienne-au-Pont, the Belgian firm built only chassis (i.e., complete running gear but without coachwork) of the highest quality. Coachwork usually was made by Vanden Plas. The illustration shows an outside-control limousine with "roof gallery" (roof rack), fitted luggage, and "French" side and rear panels on a 30/35 chassis designed by Ernst Lehmann, formerly of Daimler Motoren Gesellschaft. Four-cylinder engine, 5127 cc, 35 hp (26 kW).

Figure 55. Cyklonette, circa 1906. A cycle car built by the Cyklon Maschinenfabrik of Berlin. Despite its ungainly appearance, it was produced, with steady improvements, from 1904 to 1923. The ostensibly simple design with only three wheels and an air-cooled engine required complex engineering solutions because of its front suspension and drive. Shortly after the Cyklonette, its designer, Franz Louis Hüttel, developed a similar three-wheeler for Phänomen (Phänomobil, 1907–1927). Single-cylinder engine, 452 cc, 3.5 hp (2.6 kW).

Figure 56. Hansa 6/18 HP, 1913. A high-quality lightweight car by Westphalian auto maker Ramesol & Schmidt in Bielefeld, taken over by the Hansa Werke. At Hansa, it underwent further development under designers Nathan S. Stern and Ernst Engel, whose simplified design achieved lower weight (approximately 800 kg [1764 lb]) and a competitive price (5600 Marks). The car sported four doors and worm gear final drive (see Appendix) but was still powered by a side-valve (SV) engine. Approximately 1,000 units were built. Four-cylinder engine, 1540 cc, 18 hp (13 kW). Also built as the Hansa-Lloyd 6/20 HP.

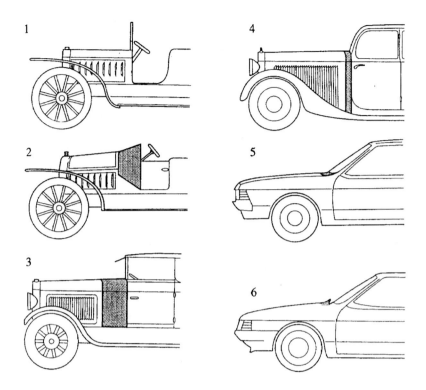

Figure 57. Evolution of the cowl. (1) No smooth transition between engine hood and upright dashboard. Adjustable windshield and leather panels available as extra-cost accessories (prior to around 1910). (2) Smooth transition (known as a torpedo, later a cowl) between engine hood and bodywork. Developed by Ludwig Kathe for the 1908 Horch Prince Henry touring cars. (3) Cowl now a separate body part, providing a mounting area for styling elements, vehicle lighting, and fresh air vents (1920s). (4) The cowl dwindles to a tiny strip between the hood and bodywork (1930s–1950s). (5) The cowl is reduced to a sheet metal or plastic component ahead of the windshield, with perforations for windshield wiper drive and heating/ventilation/air-conditioning inlets (1960s–present). (6) If the hood extends back to the windshield and its rear edge is swept upward, the cowl can be eliminated entirely. Advantages include cost reduction, better visibility, larger wiper coverage by means of "disappearing" wipers (patented by Béla Barényi in 1951), and reduction of accident damage (in common use at present).

of 1908–1910, named for the Kaiser's brother. Besides high average speeds, criteria included reliability, comfort, and economy. Foreign marques also took part in the trials; Austro-Daimler and Vauxhall both adopted "Prinz Heinrich" and "Prince Henry" as model designations. In 1908, the Ludwig Kathe coachbuilding firm developed a special sports car body for Horch's entry in the 1908 Prince Henry Trials (the Zigarre, Figure 83). The following year, several of its features, including the cowl (Figure 57) and continuous beltline, were introduced on production models.

Aside from sports cars and racers, record-setting cars played a major role in testing new engineering developments. In the United States, A.L. Riker (driving the Riker Electric) and Alexander Winton (Winton Two-Cylinder) must be credited as the originators of automotive record setting. In Europe, the role was filled by Amédée Bollée, Jr. driving his "Torpille" in 1899, and Charles Jeantaud. Around the turn of the century, Jeantaud was Europe's largest manufacturer of electric cars. In December 1898, one of his electrics was timed at 63.158 km/h (39.246 mph). Belgian racer Camille Jenatzy topped this with a speed of 66.667 km/h (41.427 mph), throwing down the gauntlet to Jeantaud. Several months of intense rivalry followed, with increasingly higher speed records. Finally, in 1899, Jenatzy, driving a streamlined electric powered by a 50 kW motor (Figure 58) reached 105.882 km/h (65.795 mph), making him the first man to break the 100-km/h (62-mph) barrier. The 200-km/h (124-mph) barrier was shattered by French driver Victor Héméry in 1909, driving a 200-hp Benz (Figure 59).

Figure 58. Jenatzy's record-setting car, 1899. Driving his cigar-shaped La Jamais Contente, *Camille Jenatzy made automotive history as the first man to exceed 100 km/h (62 mph) with a road vehicle. On April 29, 1899, he achieved a top speed of 105.8 km/h (65.8 mph) near Paris. Despite aluminum bodywork by coachbuilder Léon Auscher, his "Never Satisfied" tipped the scales at approximately 1.2 tons; its batteries alone weighed 500 kg (1102 lb). The surviving original was joined by a replica in 1997. Two electric motors (200 V, 250 amps), direct (chainless) drive to rear wheels.*

Figure 59. Blitzen-Benz, 1909. After building several smaller race cars, this vehicle, later dubbed the "Blitzen-Benz," was built in 1909, initially with 120, then 150, and finally 200 hp. Driving its most powerful variant, new records on the far side of the 200-km/h (124-mph) mark were set in succession by Victor Hémery in 1909 (202.7 km/h [126 mph]), Barney Oldfield in 1910 (211.2 km/h [131.2 mph]), and in 1911 by Bob Burman (226.7 km/h [140.9 mph]). In this photograph, Benz engineer Fritz Erle is at the wheel. Four-cylinder engine, 21,504 cc, 200 hp (147 kW).

In long-distance trials, reliability, rather than speed, was of paramount importance. After racing and transcontinental runs conquered Europe and North America, automobilists followed the example of daring cyclists and crossed Africa and Asia as well.

In 1902, the rather effusive and empathetic Otto Julius Bierbaum undertook a journey with his 8-hp Adler and became the first man to cross the Alps by automobile. Soon, in the style of the age, his feat was followed by epic long-distance expeditions that took their toll of man and machine: Paul Graetz, driving a Gaggenau, was the first to cross Africa from east to west. He departed Dar es Salaam in 1907 and arrived in Swakopmund 630 days later. Count Scipione Borghese won the 1907 Peking to Paris race with an Itala. He needed two months to cover 13,000 km (8,078 miles). An even longer race followed in 1908; *Le Matin* of Paris and *The New York Times* underwrote an automobile race from New York to Paris by way of Chicago, Seattle, Vladivostok, Irkutsk, Omsk, Moscow, and Berlin. Six months after the start, Hans Koeppen, driving a Protos (Figure 60) was the first to cross the finish line; however, he was reclassified in second place behind the American team driving the Thomas Flyer because he had loaded the Protos onto a railway car just short of Seattle.

Figure 60. Protos 17/30 HP, 1907. The apex of transcontinental motorsport was the 21,000 km (13,000 mile) New York to Paris race in 1908, contested by three French teams and one team each from Italy, Germany, and the United States. To provide weatherproof stowage for their equipment, the Protos car was given a truck body. In the course of the race, for weight reasons, this was reduced to a four-seater with a short cargo bed. Four-cylinder engine, 4560 cc, 30 hp (22 kW).

By that time, North America had developed into the world's preeminent automotive nation, leaving Europe in its wake. In 1912, Ford alone produced 170,200 passenger cars, while all British makers combined managed only 23,200. In 1913, Germany built only 12,400. In 1914, the United States numbered more than 1.3 million licensed vehicles, Britain had 245,000, France had 100,000, and Germany had only 64,000.

Commercial Vehicles and Buses to 1914

The incentive to develop gasoline-powered commercial vehicles originated in France, the world leader in road transport at the turn of the century. In 1893, Panhard & Levassor introduced a *camionnette à pétrole*, based on its Daimler-license motorcar—the world's first gasoline-fueled delivery wagon (Figure 61). With its upright front-mounted engine, change gear transmission, and Ackermann steering, it offered considerable improvement over the first German truck by the Daimler Motoren Gesellschaft, which, with its rear engine, belt drive, and turntable-and-chain steering, appeared antiquated even at its introduction in 1896 (Figure 62).

One customer for the Daimler truck was the Daimler Motor Syndicate Ltd. of Coventry. Founded by Gottlieb Daimler's business associate Frederick R. Simms in 1893, and dominated

64

Figure 61. Panhard & Levassor Camionnette, 1893. The world's first gasoline-powered delivery vehicle carried over to truck design the standard automobile layout—front-mounted upright engine, change gear transmission, rear-wheel drive, and Ackermann steering. It boasted chain instead of the belt drive still used by Benz and Daimler, good weight distribution, and, as a result, good handling. At the tiller is Emile Levassor. Front-mounted upright Daimler V2 engine, 565 cc, 2 hp (1.5 kW).

by speculator Harry J. Lawson, the concern had obtained the British rights to Daimler's patents. With the repeal of the Red Flag Act in 1896, which had inhibited any British initiative to develop mechanically driven road vehicles, a host of new opportunities presented themselves.

Initially, neither the Daimler freight trucks nor the Benz delivery vehicles, introduced to England by the firm's French distributor Emile Roger, could make any headway against British steam-powered freight wagons. Their power output was simply inadequate, their technology not yet fully developed, and their operating costs too high. Commercial vehicle trials in Versailles (1897) and Liverpool (1899 and 1901) bolstered the prevailing opinion that only steamers could move heavy loads cheaply and reliably.

Nonetheless, the first solid incentive to build gasoline commercial vehicles originated in Britain. London transportation firms felt compelled to replace their horse-drawn omnibuses with mechanically driven conveyances to remain competitive with streetcars, the Underground, surface railways, and Thames River steamers. Experience had been accumulated on steam, overhead trolley, and battery-powered omnibuses, but these did not exactly encourage substantial investment.

Für die Landwirthschaft:

Ein „**Daimler**" ist ein gutes Thier,

Zieht wie ein Ochs, du siehst's allhier;
Er frißt nichts, wenn im Stall er steht

Und sauft nur, wenn die Arbeit geht;

Er drischt und sägt und pumpt dir auch,

Wenn's Moos dir fehlt, was oft der Brauch;

Er kriegt nicht Maul= noch Klauenseuch

Und macht dir keinen dummen Streich.
Er nimmt im Zorn dich nicht aufs Horn,
Verzehrt dir nicht dein gutes Korn.
Drum kaufe nur ein solches Thier,
Dann bist versorgt du für und für.

Cannstatt
zum Volksfest 1897.

Daimler-Motoren-Gesellschaft.

Wolfgang Drück, Cannstatt.

Figure 62. Daimler truck, 1896. Folk festivals, fairs, and exhibitions were excellent venues for promoting the first motorcars. At the 1897 Cannstatter Volksfest, an unknown poet was inspired to write an ode to the Daimler truck. The Daimler Motoren Gesellschaft offered this design with 4-, 6-, 8-, and 10-hp engines, as well as payloads of 1500, 2500, 3750, and 5000 kg (up to 11,000 lb). V2 engine mounted vertically below rear floor, for example, 1530 cc, 6 hp (4 kW).

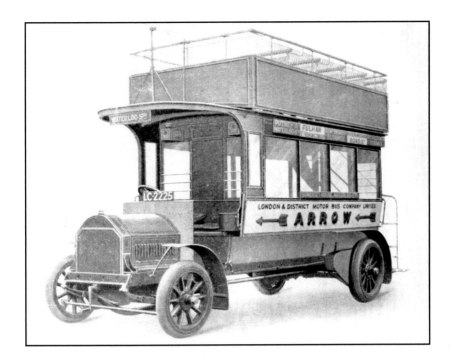

Figure 63. Stoewer Omnibus OS I, 1906. The Stoewer Works of Stettin had estab-lished commercial ties with England as early as 1902. In 1905, the firm received a major order from London for 200 buses (presumably only rolling chassis). Typical for the time were the so-called "impériale" (upper deck), open driver compartment, rear entry, and solid rubber tires. Four-cylinder engine, 5880 cc, 30 hp (22 kW).

Beginning in 1905, continental firms benefited from Lawson's paranoia and eagerness to hound any motorcar manufacturer who did not pay royalties to him, and the inability of British vehicle manufacturers to build halfway usable motor omnibuses even well after the repeal of the Red Flag Act. The main beneficiaries were German makers: Daimler and Büssing granted licenses, while Stoewer (Figure 63), Dürkopp, Scheibler, and Argus delivered truck chassis directly to English bus companies that installed locally made omnibus bodies. The newly born German commercial vehicle industry relegated its competitors from France, England, Belgium, and Switzerland to also-rans, and henceforth represented an interesting economic segment with great export potential for the State (and the tax collector).

In Germany itself, the beginnings of bus traffic proceeded at a pace no less ponderous than in Britain. Bus lines founded in Siegerland in 1895 with two Benz omnibuses (Figure 64), and further lines elsewhere with Daimler and Benz vehicles from 1898, were soon abandoned; their maintenance and repair costs exceeded revenues. Aside from technical difficulties, rutted roads that became virtually impassable in bad weather made scheduled passenger ser-vice practically impossible.

Figure 64. Benz motor bus, 1895. As a predecessor of the later Siegtalbahn (Sieg Valley Railroad), 1895 saw the establishment of the world's first bus line, serving the Siegen area. The small bus was based on the 1893 Viktoria (Figure 33). The enclosed passenger compartment could carry six persons. Before the year ended, operational problems forced Benz to take back this and a second vehicle. Single-cylinder engine mounted horizontally at the rear, 2650 cc, 5 hp (3.7 kW).

The breakthrough was achieved by Heinrich Büssing (1843–1929), who built a special plant for trucks and omnibuses in 1903 (Figure 65). In 1904, to stimulate lagging demand, he established a mail-carrying bus line between Braunschweig and Wendeburg. In 1908, he founded a "transport company for the conveyance of wares and goods" in Berlin, making Büssing a pioneer of both the bus and motor freight industries. As of 1905, his firm was second only to Daimler in delivering omnibuses to the forerunners of today's London Transport, and shortly thereafter to the ABOAG.

The ABOAG—the Allgemeine Berliner Omnibus-Aktien-Gesellschaft (Berlin General Omnibus Corporation)—inaugurated its service in November 1905 using Daimler buses, followed by double-deckers by Gaggenau and NAG. In 1906, the Compagnie Générale des Omnibus (CGO) introduced bus service to Paris, with vehicles by Eugène Brillié (Figure 66). The first German postal motor coach line was established between Bad Tölz and Lenggries in 1905, using Daimler buses. Smaller cities such as Darmstadt and Baden-Baden also initiated motorized bus service around this time.

Figure 65. Büssing, 1905. This advertisement in an automotive publication illustrates the state of the art around 1905: iron-shod wheels for trucks, solid rubber for buses, in both cases open driver cabs, identical chassis, and chain drive between the front-mounted engine and rear-drive wheels. The bus has a rear entry, similar to a streetcar, and a luggage rack (with access ladder) above the driver. Four-cylinder engine, 4760 cc, 20 hp (15 kW).

Figure 66. Brillié-Schneider P2, 1906. Following a call for bids, the Parisian Compagnie Générale des Omnibus—the CGO—selected a bus designed by Eugène Brillié and built by armaments maker Schneider. Special features included the driver position high above the engine, as well as the "impériale," in this case an upper deck fitted with a front panel and roof (i.e., a transitional design between the open upper deck and the enclosed double-decker). Four-cylinder engine, 6650 cc, 35 hp (26 kW).

The motor buses of those early efforts were not technically distinctive products. Their chassis originated in truck practice, and their bodywork was derived from that of horse-drawn buses or, in some cases, railway cars. From around 1903, they rolled on steel wheels, covered with solid rubber tires. Power output of their two- or four-cylinder engines amounted to approximately 30 hp and required gearing that limited top speed to approximately 20 km/h (12 mph).

The demand for trucks was even more limited than that for buses. Purchase price and maintenance costs were high, prompting trades and industries to retain their horse-drawn transport. Regardless, the preconditions for motorized truck traffic had already been firmly established—thanks to railroads, the distances between factory and market and between workplace and home had grown. These distances could be conquered only by horse and wagon under special circumstances, as the time factor began to assume an increasingly important role.

Beginning in 1903, Lt. Col. Edmund Troost commissioned NAG to build tractor-trailers to a design by engineer Joseph Vollmer. Two of these were sent to Namibia (which at the time was German Southwest Africa) and placed in service with the local colonial troops. These so-called DURCH trailer trains, rolling on iron tires (Figure 67) could move 18,144 kg (20 tons) of payload. Propelled by 45-hp engines, their top speed was 10 km/h (6.2 mph).

In 1908, Siemens-Schuckert engineer Wilhelm A. Th. Müller-Neuhaus developed his "Müller Train" using a tow vehicle and six steerable trailers (Figure 68). A gasoline engine drove a DC

Figure 67. NAG DURCH road train, 1903. After efforts with an imported steam traction engine of 1896 ended in failure, the DURCH train represented Lt. Col. Edmund Troost's second attempt at motorizing colonial troops in German Southwest Africa (now Namibia). Because of the heavy load of two trailers, the rear wheels of the tractor were not driven by chain but rather by gear and pinion. Four-cylinder gasoline/alcohol engine, approximately 11,000 cc, 45 hp (33 kW).

Figure 68. Müllerzug, 1908. The Müller Train was more capable than the DURCH in rough terrain but was also more complicated. All wheels of the train were driven. The lashup consisted of a double-ended tractor carrying two gasoline engines and generators, plus six trailers, each with two standardized pivoting axle assemblies. These were joined by a mechanical "anti-parallel" linkage to ensure precise tracking, plus a cable for electrical power transmission. The train could carry a payload of 30 tons (DURCH: 10.7 tons). Two 60-hp gasoline engines and two 75 kW generators.

generator, which in turn fed power through cables strung between the trailers, to power electric motors mounted in their wheel hubs. The Müller Train could move 30 tons of payload and inspired Ferdinand Porsche's Austro-Daimler Landwehr and C-Train designs. The culmination of early road train development was the "Australzug" ("Australtrain") developed in 1913 under contract for a firm in Melbourne. A predecessor of the Australian road trains of today, it had ten trailers and a total payload of 30 tons. All of its trailer wheels were electrically driven.

These unique, exotic vehicles could not disguise the fact that German interest in trucks was virtually non-existent, with the result that leadership in commercial vehicles passed to England and France shortly after the turn of the century. A German Reich survey first conducted in 1907 indicated that of 27,026 motor vehicles of all types, only 1,211 or 4.5 percent were used in freight service. This figure also included motorcycles with package racks (254 vehicles) and delivery vehicles derived from passenger cars, making it impossible to give an exact figure for how many "real" trucks actually roamed the roads.

Thanks to government subsidies, by 1911 the proportion of motorized trucks rose to 7.5 percent of all motor vehicles. Beginning in 1908, the German military subsidized purchase and maintenance costs of trucks used in civilian businesses, under the stipulation that the vehicles be maintained in usable condition, were presented for annual inspection, and could be requisitioned for military service in time of national emergency. This largely eliminated the arguments against higher operating costs. Breweries, grain mills, brickworks, and other businesses gradually began to motorize their fleets.

The Role of Motorized Vehicles in World War I, 1914 to 1918

On land, World War I began as the Franco-Prussian War of 1870–1871 had ended: on foot and on horseback. Cavalry, which had been an important arm in the eighteenth and early nineteenth centuries and in 1815 played a decisive role at the battle of Waterloo, had diminished in numbers relative to infantry. However, military planners regarded successful mounted campaigns of the 1870–1871 war as proof of their military effectiveness and carried their experiences into the twentieth century. According to the planners, infantry, artillery, and cavalry would decide the battles of World War I as well. The strategists of the combatant nations saw the motor vehicle, the most mobile product of modern technology, as only of limited assistance in armed confrontations.

Nevertheless, the German High Command (Oberste Heeresleitung—OHL) saw trucks as suitable for the transport of munitions, provisions, and "military materiel—[and] as a link between the railway and horse-drawn transport which can haul smaller quantities across unfavorable terrain and cross country" (Bürner, R., *Der Kraftwagen im Felde,* "The Motorcar in the Field," Berlin, ca. 1916). Lacking all-wheel drive, no more could be expected of the German truck (Figure 69).

Figure 69. Nacke subsidy truck, 1916. Subsidy guidelines established by German Army planners represented a windfall for existing manufacturers and prompted outside firms to take up truck manufacturing. Before the outbreak of war in 1914, approximately 15 firms were approved to build subsidy trucks, including the Automobilfabrik Nacke (Nacke Automobile Works) in Saxony, which had begun building delivery trucks and small buses in 1902. This advertisement is dated 1914.

Because the German army could not bear the expense of a large motor pool, which moreover would have to be constantly modernized (in contrast to modern practice), it established guidelines for the design and construction of trucks. These guidelines included 6000 kg payload (with trailer), engine output initially 30 hp, after 1913 35 hp, maximum speed 16 km/h (10 mph) on solid rubber tires, and range at least 250 km (155 miles) on a single tank of fuel. For civilian firms, acquisition of trucks meeting these requirements was subsidized at 5000 Marks and later 4000 Marks. Annual operating costs were offset by 1000 Marks per year for five years. Other nations, such as Great Britain, had similar systems.

Despite state subsidies, it is unlikely that German annual truck production exceeded 3,000 units before 1914, far too few in the event of war with other industrialized nations. When the High Command became aware of the problem, it increased production of "subsidy trucks" to approximately 15,000 units in 1917. This benefited not only the existing plants but also firms well removed from the industry, which took up truck production as subcontractors. These firms included Hille in 1911, MAN and Vomag in 1915, and Magirus in 1916.

The trains designed by Porsche, tractor-trailer sets that could operate equally well on railroad tracks or roads, assumed special significance for the Austro-Hungarian army. Tractors (in the automotive sense) are vehicles related to trucks, although they are directly derived from them. In the Crimean War of 1854–1856, the British made use of a traction engine originally intended for peaceful purposes. In 1870, the German Reich bought two steam traction engines from the English firm of John Fowler, but these were not particularly successful. Traction engines faded into obscurity until the gasoline engine opened new possibilities. While Porsche's gasoline-electric (hybrid) trains were successfully applied to artillery transport (Figure 70), the smaller gasoline tractors offered by Daimler, Büssing, Dürkopp, and Skoda found little military interest.

Armored vehicles are older than the automobile itself. Assyrians and Egyptians had movable fortresses, whereas the Greeks, Romans, and Chinese employed war chariots. In the Middle Ages, muscle- or even wind-powered engines of war appeared—war wagons, land ships, and rolling siege towers. The first steam-powered armored vehicles appeared around 1850. All of these power sources, however, proved inadequate.

Figure 70. Austro-Daimler Lasttrain, 1914. The AD Load Train represented a continuation of the Müller Train; mounting flanged wheels allowed it to travel on rails as well. The tractor mounted a hybrid drive, with the individual axles of the trailers electrically driven. It was employed to tow mortars (as shown here) or up to 10 trailers. Payload was as much as 70 tons. 90 hp (66 kW) or 150 hp (110 kW) gasoline engine driving electrical generator.

The birth of the gasoline engine promised more rewarding avenues of development. Opel and Ehrhardt (Figure 71) in Germany, Minerva in Belgium, Charron, Girardot & Voigt in France, as well as Rolls-Royce and Lanchester in England, to name only a few, turned production passenger cars into armored cars. Because of their increased weight and single-axle drive, the usefulness of these vehicles in rough country must have been limited. Paul Daimler, son of Gottlieb and up to 1905 technical director of the Austro-Daimler Motoren Gesellschaft (Austro-Daimler Motor Company), designed one of the few passenger-car-based armored cars to be equipped with all-wheel drive. A decade later, Ehrhardt, Büssing, and Daimler demonstrated prototypes on truck chassis. Few of these were actually produced.

Of far greater significance for the course of the war were the armored vehicles called "tanks." Compared to wheeled vehicles, they were much more suited to off-road use because their tracks had much lower ground pressure and therefore

Figure 71. Ehrhardt armored car, 1906. As early as 1906, Heinrich Erhardt, founder of what would later become Rheinmetall, displayed a Kanonen-Automobil—"a cannon car"—at the Berlin International Automobile Exposition. This consisted of a passenger car chassis fitted with an armored body and anti-balloon cannon. This was the predecessor of the armored scout cars employed for the most part by British and Belgian forces in World War I. Four-cylinder engine, 9230 cc, 45 hp (33 kW).

did not sink as deeply into soft ground. In addition, and in contrast to the first road-going armored cars, their overall layout, subsystems, and engine output were better suited for their intended purpose.

The term "tank" arose from the designation "Watertank for Petersburg," which the British used to camouflage the true nature of the vehicles. Designs for tanks, assault cars, armored land cruisers, *chars d'assaut*, or self-propelled guns had already been presented, in some cases before the war, by Burstyn, Goebel, Corry, Swinton, Bremer, and others. They were based on the well-known linked-track tractors, primarily built as of 1901 by American firms such as Lombard, Holt (formerly Hornsby), Best, Monarch, and Keystone for the lumber, farming, and construction industries. The tank as a combat vehicle equipped with linked tracks therefore is not a wartime invention but rather a linked-track prime mover or tractor adapted to military duty. Because tanks do not represent a means of transportation, they will not be examined in greater detail here.

After the Armistice, both victor and vanquished agreed that the armored vehicle had decided the outcome of the war, along with the entry of the United States. Even before its entry into the war in April 1917, the American war effort considerably increased the already superior material, financial, and numerical superiority of the Allies. As early as 1914, the French general staff ordered more than 1,000 trucks from American manufacturers, 600 from White alone. Beginning in 1910, the former maker of steam cars offered gasoline-powered trucks as a licensee of the French Delahaye firm and during the war delivered thousands of trucks to the U.S. Army.

Similarly to the German "standard truck" built to meet the subsidy requirements, in the United States the Quartermaster Corps, responsible for military vehicles, issued specifications in March 1917 and engaged approximately 50 engineers from the commercial vehicle industry to develop and produce a 1.5-ton (Class A) truck and a 3-ton (Class B) truck.

The first two trucks were ready as early as October 1917, a feat that was accomplished only by distribution of the work and use of proven components. The four-cylinder engine of the three-tonner, for example, was made by Continental (cylinder block), Waukesha (crankcase and cylinder heads), Hercules (pistons), and other suppliers. Similar outsourcing was applied to other components such as the transmission, front and rear axles, brakes, and steering.

Fifteen different truck manufacturers rapidly assembled the interchangeable parts and sub-assemblies to create the Army Truck, which appeared mainly on the Western Front as the Liberty (or USA). After the war, it was used by Willème in France, among others, as the starting point for their own truck production. A total of 9,452 examples of the three-tonner (Figure 72) were built; the ton-and-a-half was less significant.

The less than overwhelming number of Liberty trucks (production was halted by the U.S. government at the end of the war) was greatly exceeded by other manufacturers producing their own designs. White alone delivered 18,000 trucks; GMC produced 16,000 two-tonners and 5,000 ambulances. Other, smaller firms built a total of 10,000 to 20,000 units. Added to these were approximately 11,500 Nash Quads and 15,000 FWD trucks, both with four-wheel drive and four-wheel brakes; the Nash also had four-wheel steering. With the all-wheel drives, the Allies for the first time had an all-terrain truck that could not only ensure supply on solid underfooting but could also travel cross country to support forward positions.

World War I, the watershed between antique and modern warfare, between saber-rattling cavalry and motorized combat units, represented the end of muscle-powered war and transport equipment. It provided a foretaste of fully mechanized hostilities, not only on land but also on the sea and in the air.

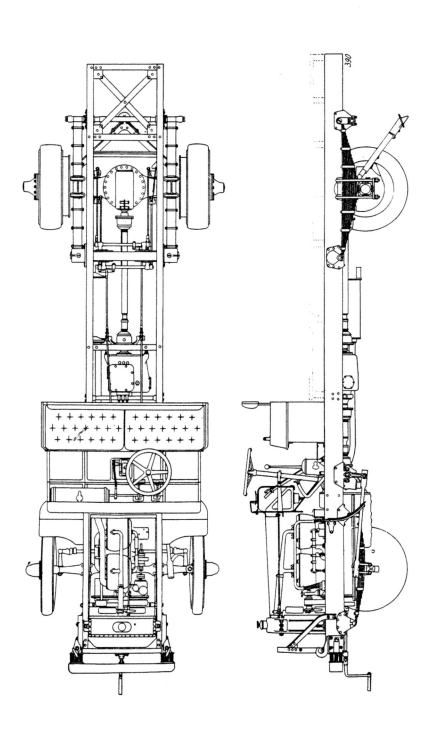

Figure 72. Liberty Class B truck (USA), 1917. The layout of the Army truck represented the American state of the art at the time: straight ladder frame with transverse braces, three-point engine mount and separate transmission, engine equipped with magneto and battery ignition, and fuel tank mounted on the instrument panel. To achieve the required ground clearance, the worm gear is mounted above the differential; for convoy operations, the radiator and vehicle front are protected by a grille and spring-mounted bumper. Four-cylinder engine, 6950 cc, 58 hp (43 kW).

A view of things to come was provided not only by tanks but also by the endless columns of Allied trucks bringing supplies and reinforcements, dispatch riders on motorcycles as a contemporary replacement for horses, motorcycle infantry replacing mounted troops, and the use of conventional passenger cars. This culminated in September 1914, when 600 Paris taxis were requisitioned to move 6,000 troops to the Battle of the Marne. An unknown driver summed it up in an article in the German magazine *Der Motor* (April–May 1915, p. 54): "The whole secret of the war seems to be speed."

The Automobile as an
Industrial Product
1919 to 1945

American Superiority

World War I left in its wake a new economic, political, and social constellation. The United States, once a debtor nation, had been transformed into the world's largest creditor. On the other hand, the nations of Europe, even the victors, were economically devastated. In those nations where war had brought development to a standstill, automobile production resumed haltingly at best. Demand eventually rose, but it was no longer sufficient to appeal to nobility with hand-built, expensive coachwork. Instead, emphasis shifted to motorizing the newly arisen middle class, however limited its purchasing power.

European manufacturers were poorly equipped to serve this new class of customer. The dominant philosophy of the time was the same as that expressed in 1915 by a leading auto maker: "…despite the best and most modern manufacturing facilities, it is not possible for three workers to build more than two Daimler cars per year without neglecting the absolutely necessary hand craftsmanship, as for example is done in America." (*Zum 25jährigen Bestehen der Daimler Motoren Gesellschaft*, "On the 25th Anniversary of the Daimler Motor Company," Untertürkheim 1915, p. 129).

Although Europeans considered hand craftsmanship absolutely indispensable, the post-World War I years saw the completion of the transformation of American industry from craftsmanship to industrial mass production, as epitomized by Henry Ford. No longer did the skill of specialized, highly paid craftsmen in the factory halls determine the quality of the final product, but rather the systematic and logical work of engineers responsible for the design of the product and its manufacturing process.

The key to the success of American auto plants lay in a shrewd and purely economic perspective to which all else was subordinate. While an American engineer designed a component to be as good as necessary, his European counterpart made it as good as possible and attractively shaped as well. The American industry paid attention to manufacturing the part with as little investment of time and material as possible—if the part was not already available from an outside supplier in the first place.

In the United States, the supplier industry was already well established. While the advertising literature of European auto makers proudly proclaimed that most parts were made in-house, many American plants bought axles, fenders, radiators, rims, steering, transmissions, and even entire engines from specialist suppliers that could produce in volume and therefore offer attractive pricing.

The American car dealer did not regard a signed sales contract as the end of his efforts; rather, he viewed it as the beginning of a partnership between the factory and the customer. Larger manufacturers offered repair stations, fixed replacement parts pricing, and rebuilt components, services that were not yet known in Europe.

American business practices began to represent a threat to European markets, particularly after 1925 as the U.S. market became temporarily saturated and excess production had to be sold, one way or another. American auto makers not only flooded the markets of European auto-making nations, but also expanded into the vacuum that the war-strapped European industrialized nations had left in their wake on the Iberian Peninsula, Scandinavia, and Eastern Europe.

American firms not only exported finished products, but also erected assembly plants in England and Germany. Henry Ford made the first foray, in Manchester, in 1911. After only eight years, he had achieved a market penetration of 40 percent of all vehicles registered in England. Willys-Overland followed in 1921, likewise in Manchester. Four years later, General Motors expanded into the European market. Since 1919, Ford's most serious rival, General Motors, bought Vauxhall in England and built an assembly plant in Hamburg in 1925. Two years later, General Motors opened a second German plant in Berlin. Chevrolet passenger cars and trucks rolled from these assembly lines until General Motors bought the Opel Works in Rüsselsheim, near Frankfurt, in 1929.

Until the onset of the worldwide economic depression (1929–1932), Chrysler, Hudson (Figure 73), Willys-Overland, and Studebaker also assembled cars in Germany. In 1925–1926, Ford assembled cars in Berlin, then built a permanent plant in Cologne in 1930–1931. The portion of imported cars and cars assembled from foreign components (Fiat and Citroën also operated assembly plants in Germany) reached its peak in 1929, with 40 percent of new German car registrations.

Added to business methods unfamiliar in Europe and the sound financial backing of the American auto industry, the automobile soon enjoyed a degree of everyday utility, born not so much out of a desire for technical progress (Figure 74) as out of practical considerations. For example, Edward Gowen Budd had concluded from numerous railway accidents that wood was in many cases an inadequate construction material, and he replaced it with sheet metal. He also replaced

Figure 73. Essex Super Six, 1928. The president of Hudson Motor Car Co., Roy D. Chapin, saw a second model line as a means to increase sales and profits for the firm. Indeed, in 1919, only a year after its introduction, the company's Essex nameplate outsold Hudson. The Essex developed into a hot seller not only in the United States, but also in Britain, Scandinavia, and Germany (note contemporary family snapshot) and elevated Hudson into one of the leading auto makers of the 1920s. Six-cylinder engine, 2511 cc, approximately 45 hp (33 kW).

the wooden framework of automobile bodywork with stamped steel members. Budd's first large customer was Dodge, which in 1914 ordered 5,000 all-steel bodies—because Budd could offer them for $10 less per unit. Shorter assembly time and a simpler paint process were added advantages. Increased torsional stiffness and, for enclosed bodywork, added rollover protection were completely incidental benefits.

Until the mid-1920s, auto makers regarded the skipping, bouncing chassis as a rigid foundation for engine and transmission, which were solidly bolted to the frame. The results were deformations, noise, and, not uncommonly, mechanical failure. Maxwell addressed the problem in 1925, and Oakland in 1930, with drivetrain components mounted on steel springs. The most convincing solution was achieved by Chrysler and Firestone. Their so-called "Floating Power" engine mounts (Figure 75) consisted of flexible rubber blocks vulcanized over metal hardware. Citroën acquired licenses for France, while Mecano GmbH of Frankfurt did the same in Germany. In short order, Continental, Phoenix, and Getefo followed with rubber/metal mounts of their own, which have remained part of automotive technology since that era.

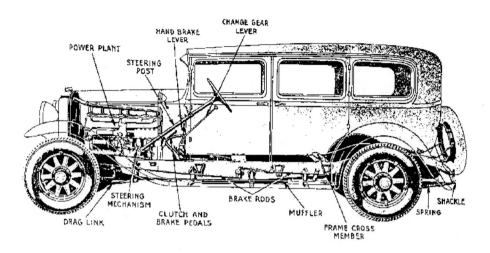

Figure 74. American Standard, circa 1928. The standard layout—with its ladder frame, water-cooled inline engine, rear-wheel drive, and longitudinal leaf springs front and rear—was an obstacle to development of lower, safer, and faster cars. Only detail improvements could be expected of the standard design, such as balloon tires (here, with an aspect ratio of 98 percent), steel disc instead of wooden spoked wheels, and hydraulic instead of the mechanical brakes shown here.

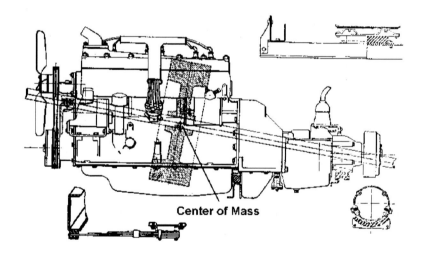

Figure 75. Chrysler Floating Power engine mounts, 1931. Two rubber and steel elements are attached to the engine and transmission in such a way that the imaginary line through them also passes through the center of mass of the engine/transmission unit. Combined with a leaf spring mounted below the bell housing, these reduced noise and torque reaction during engine load changes. Applied to Plymouth in 1931, Chrysler in 1932, and licensed throughout the world.

The 1920s were the golden days of the U.S. auto industry. Never again would American car makers achieve such high export rates and have so much influence on automotive technology. Later American cars were less ambitious from an engineering standpoint, and they were too large and too expensive for most world markets.

The Situation in Germany, 1918 to 1932

Until early 1933, a succession of German governments failed to recognize the economic impact of the automobile industry. For too long, the passenger car was regarded as an essentially superfluous luxury, which was reflected in regulatory actions against the automobile and its environment. Government bureaucrats and elected officials, still mired in the age of the railway, saw the passenger car, and especially the commercial vehicle, as an unwelcome competitor for steel-rail transportation systems. A sharing of labor between road and rail was, then as now, far beyond the horizon.

Consequently, from the beginning, various German state governments collected fees from the automobile. Actual vehicle taxes were preceded by "stamp duties." For example, in Hesse from 1899 and Lübeck from 1902, fees of 5 to 50 Marks were levied, depending on the size, list price, and horsepower of a motorcar. Beginning with the tax law of July 1, 1906, passenger cars were assessed an annual luxury tax: 25 Marks for cars less than 6 hp, and 50 Marks for cars less than 10 hp. Vehicles used for commercial transportation of people and goods were exempt from this tax. The tax rate was based on so-called "taxable horsepower," which was less than actual engine output. These figures were reflected in automobile type designations, for example, the Protos 17/30 HP (17 taxable, 30 actual horsepower, Figure 60).

The concept of general taxation was not established until the post-World War I tax reform. The motor vehicle tax law of April 1922 imposed a tax on all vehicles used to transport people or goods. A car under 6 (taxable) hp was taxed at 100 Marks, under 10 hp at 200 Marks, and under 12 hp at 300 Marks per year. The only exception was cars under 8 hp used by physicians. This led to auto makers regularly offering so-called *Doktorwagen* in their model ranges. At the same time, several unlawful but nevertheless lucrative taxes imposed by individual states were eliminated, such as the Berlin tire tax, Bavarian pavement toll, and the local vehicle taxes of communities such as Munich, Dresden, and Hildesheim.

Other administrative measures also contributed to making car ownership more expensive. On top of heavily taxed mineral oil products, 1909 saw the introduction of mandatory liability insurance. Clearly visible license plates were legally required as of 1901. Beginning in 1910, more stringent driver's license testing requirements replaced the earlier police certificates.

During World War I and in the early postwar years, German automotive technical progress stagnated. American and, to a lesser extent, British automobile and component manufacturers had taken the lead in technology, production, and marketing. If German firms wanted to catch up to the international standard, they would have to adopt foreign products and methods, and accede to paying licensing fees.

With the introduction of the Rentenmark in November 1923 and reorganization of war reparations, Germany enjoyed a period of stabilization in the automobile sector and beyond. Thanks to high bank interest rates and low wages, foreign investment streamed into Germany. The industry was able to rebuild and modernize. Many new auto manufacturers sprang up, all wanting their share of the economic recovery:

- In 1922, 46 car manufacturers produced 90 different models.
- In 1923, 63 car manufacturers produced 108 different models.
- In 1924, 86 car manufacturers produced 146 different models.

The pace of economic recovery came as a surprise, especially in view of high protective tariffs imposed in 1918. On the basis of a misguided policy of liberalization, Germany embarked on an "unholy era of customs policy, with its accompanying tariffs, without mutual concessions on the part of the other nations" (Robert Allmers, chairman of the RDA—the German Automobile Industry Association, Reichsverband der Automobilindustrie—in *Kraftfahrt tut not,* "Motoring Is in Need," Berlin, 1933, p. 3). In 1925, import restrictions were eliminated, and beginning in 1926, the weight duty was dropped in stages from 250 Reichsmarks to 75 Reichsmarks per 100 kg (220.5 lb) of vehicle weight. According to the *Berliner Tageblatt,* as of July 1928 after conversion of weight duty into valuation duty, the equivalent duty had dropped to 28 percent, markedly lower than the customs duty of the major European auto-making nations:

England	33-1/3 percent
Italy, Czechoslovakia	45 percent
Austria	47 percent
Belgium	65 percent
France	180 percent

Only Canada, at 20 percent, and the United States, with 25 percent, were significant exceptions; however, because of higher freight costs, only the absolute premium class of German automobiles (e.g., Maybach, Mercedes, and Benz) or special-purpose vehicles (e.g., Magirus fire engines and Krupp streetsweepers) were exported to North America.

Because of low import duties, the technical superiority of foreign competitors, and the expensive production methods of domestic auto manufacturers, imports (especially American cars) flooded the German market. "The American Peril" was a catch phrase of the time. German manufacturers appealed to national pride ("Germans buy German") or insinuations that American cars were cheap mass-produced trash. The new Adler Standard 6 (Figure 80) and Horch 8 (Figure 82) were much better answers to foreign competition than patriotic appeals or symbols such as the acorn, meant to represent German craftsmanship. However, even these cars could not stop the flow of imports. The upturn in the domestic automotive industry ground to a halt in 1928; the multiplicity of German firms dwindled from 86 in 1924 to 19 in 1928. Soon, the onset of the world economic crisis made matters even worse. Production declined rapidly (Table 4).

TABLE 4
MOTOR VEHICLE PRODUCTION IN GERMANY
1926 TO 1932

	Cars	Trucks/Buses	Motorcycles
1926	31,958	5,211	47,477
1927	84,668	11,972	81,698
1928	101,701	20,960	160,782
1929	92,025	16,230	195,686
1930	71,960	9,985	98,574
1931	58,774	8,734	51,148
1932	41,727	4,509	36,262

RDA—the German Automobile Industry Association, Reichsverband der Automobilindustrie, 1939, p. 11.

In addition, a loophole in the customs laws led to even greater foreign penetration in motor vehicle registration numbers. The Reich revenue service imposed a duty of only 10 percent on finished parts that could be installed on engines or automobiles. This prompted Ford, General Motors, and Citroën, among others, to establish assembly plants in Germany. The significance of these plants diminished after January 15, 1928, when the same duties were applied to individual parts as had been imposed on completed vehicles. The Ford and Citroën assembly plants gradually switched to German suppliers, bought domestic raw materials, and developed into independent manufacturing plants. In 1929, General Motors closed its own assembly plant and instead acquired Opel.

Compared to Ford, the action of General Motors proved to be a clever strategic move. By this time, customers had realized that the operating costs of American cars, with their large and thirsty engines and their high spare-parts costs, were too high for most German and European drivers. At first, Ford did not draw the logical conclusions and until 1933 continued to build American-style cars with engines from two to four liters. However, Opel was considered a manufacturer of economical cars matched to German conditions. To the great delight of General Motors, Opel from the beginning produced more cars than Ford Germany.

Because of pressure from imports, the existing installed base of cars, undercapitalization of German auto plants, and the large number of competing makes in a limited market, the German auto industry underwent a significant structural transformation.

Two sales associations had been formed in 1919: (1) the Deutsche Automobil-Konzern (DAK), encompassing the firms Dux, Magirus, Vomag, and Presto; and (2) the Gemeinschaft Deutscher Automobilfabriken (Association of German Automobile Manufacturers, or GDA), with NAG, Hansa/Hansa-Lloyd, and Brennabor. These dissolved in 1926 and 1929, respectively, in the wake of their inability to reconcile product planning of dissimilar lines produced by their member firms. Only two large automotive concerns remained: (1) Daimler-Benz AG, formed in 1926 by the fusion of the Daimler Motoren Gesellschaft with Benz & Cie; and (2) the short-lived Schapiro Group, consisting of NSU, the Cyklon Maschinenfabrik, the Gothaer Waggonfabrik, and several supplier firms. The formation of a German auto trust, which many hoped would provide an economic advantage, failed due to disagreement among banks as to the value of individual stocks.

German firms tended to follow their own (often eccentric) paths. This hindered formation of consortiums, and with the firms' misjudging of the market and overestimating their own abilities, led to a basically necessary and desirable process of elimination. Fafnir, Apollo, Dürkopp, Ley, and Cyklon abandoned automobile production. Other firms such as Röhr, Stoewer, Hanomag, and Brennabor filed for bankruptcy. In 1928, Dixi was acquired by BMW. The American firm Edward G. Budd Manufacturing Company bought body manufacturers Ambi of Berlin and Lindner of Ammendorf. These supplied steel bodies and stampings to Adler, Hanomag, Dixi, BMW, and Ford. In 1929, Fiat SpA founded NSU Automobil AG in Neckarsulm to assemble its cars under the NSU-Fiat brand. Through his Zschopauer Motorenwerke (DKW), Danish entrepreneur Jörgen Skafte Rasmussen exerted his influence on Audi. Under pressure from the State Bank of Saxony (Sächsische Staatsbank), the Auto Union AG was founded in 1932, combining the firms DKW, Audi, Horch, and the automobile activities of the Wanderer-Werke. Of the roughly 130 nameplates existing in 1921, only approximately ten survived the 1920s.

Germany was hit particularly hard by the world economic crisis. Short-term foreign credits, which had been put into long-term investments and financed reconstruction from 1923 onward, were called in. Production and sales suffered drastic cuts, while unemployment figures skyrocketed to more than 6 million in the winter of 1932–1933. The number of auto industry workers dropped from approximately 90,000 in 1928 to roughly 30,000 in 1932. Because of non-renewals, the total number of registered vehicles dropped by 34,000. Naturally, the percentage of imports also dropped—from 32.7 percent in 1928 to 12.3 percent in 1932. Germany had evaded the "American Peril."

In 1930, to "stimulate the economy," the German Reichstag's taxation committee and the Reich finance ministry could think of nothing better than to impose mineral oil taxes and introduce so-called *Spritbeimischungszwang*—"mandatory fuel blending." This involved government support, at the expense of automobile owners, for the total of 5,161 potato distilleries among the 5 million agricultural businesses within Germany (1931). At 796 Reichsmarks, the average tax burden per German automobile was the highest in the world; significantly, the United States, with the equivalent of 128 Reichsmarks in annual taxes, had the lowest. In July 1932, which is normally the peak of the travel season, 132,000 German owners canceled their vehicle registrations, which cut vehicle tax revenues by 20 percent.

In April 1928, the weekly standard wage of a skilled worker in the "means of production" sector earned an average of 51.84 Reichsmarks, or 207.36 Reichsmarks per month. In October 1927, the average monthly salary of a government employee ranged from 177.75 to

933.58 Reichsmarks, from an unmarried worker at the lowest salary level to a top earner in the highest bracket, married with two children. The ordinary worker would have to save for approximately ten months to be able to afford minimal motorized transport (Hanomag "Kommissbrot" [Breadloaf], single-cylinder, 10 hp, 1975 Reichsmarks). The single government employee needed eleven months to do this, and the top earner somewhat longer than two months to afford the same car.

After 40 years of auto making, the purchase of a new car seemed to be within reach, compared to the years before World War I, in which a worker had to save for three years to afford a small car. However, because the worker still had to pay rent and care for his family, in the 1920s a personal car remained a dream for most. Nonetheless, many made the "social climb" from bicycle to motorcycle, the "automobile of the common man." In 1928, a 250-cc Zündapp with a 6.5-hp engine cost 860 Marks, or approximately four to five months' wages. By comparison, for the same three to four months of his time, an American Ford worker could buy a full-fledged car, a Model T (known as the "Tin Lizzy") powered by a four-cylinder 24-hp engine.

Europe Between Imitation and Independence

Europe's prewar leadership in automotive technology had passed to the United States. It was time for the Europeans to pass the torch. They should have felt no shame in recognizing American superiority and should have tried to learn from these new transatlantic methods. However, as far as automotive development was concerned, Europe initially rested on its laurels.

Among the few Europeans who switched back from their erstwhile roles of master and again became apprentices were André Citroën, Wilhelm von Opel, William Morris, Herbert Austin, and Giovanni Agnelli. All had visited the United States, toured Ford's plants, and returned home convinced that assembly line production of automobiles should be possible in Europe as well.

Citroën dissected and studied an American car, imported expressly for this purpose. He charged his chief engineer Jules Salomon with the design of a lightweight car, which commenced delivery in June 1919. The resulting Type A had left-hand steering (unusual at the time), a complete electrical system, and demountable disc wheels. It could be equipped with any of five different bodies at the factory—a novelty at the time (Figure 76). With the Citroën Type A, Europe had its first mass-produced car developed and built using American methods—only six years after Ford's breakthrough. A total of 24,093 examples rolled off the assembly line.

In planning his own assembly line product, Opel also drew on the work of others—Citroën. In 1922, Salomon developed the 5 CV, which became one of the most popular small cars of the 1920s. Nicknamed *Citron* (Lemon) for its most common yellow color, Opel built 16,735 exact copies beginning in 1924—in green, and without paying Citroën any license fees. The color soon earned the car the nickname of *Laubfrosch* (Tree Frog) (Figure 77). With the

Figure 76. Citroën A assembly line, 1919. The tiny Citroën A was the first European car built on an assembly line, following American practice. Admittedly, the assembly line in war-torn Europe initially consisted of wooden rather than steel rails, and the cars were pushed by workers rather than being pulled by chains. Four-cylinder engine, 1327 cc, 18 hp (13 kW).

Figure 77. Opel 4/12 HP "Laubfrosch," 1924. Assembly line production was introduced to Germany by the Opel 4/12 HP "Tree Frog." It formed the initial model of an entire line of 4-hp models that lasted until 1931; almost 120,000 examples were built. At the conclusion of a 1925 auto and motorcycle race on Opel's break-in track in Rüsselsheim (opened in 1919), spectators were treated to a review of a full day's production of 125 Tree Frogs. Four-cylinder engine, 951 cc, 12 hp (8.8 kW).

4/12 HP, as the Frog was officially known, Opel not only introduced assembly line production to Germany but also took a page from Ford's American notebook and established competent repair service and fixed prices for repair parts.

Citroën and Opel aside, the Austin Seven became the most successful European mass-produced car of the 1920s (Figure 78). During its 18-year production run (1922–1939) it reached a total of 291,000 examples. The Seven, also known as the Austin Baby, was built under license in France (Rosengart), Germany (Dixi/BMW), and in the United States (American Austin/American Bantam). In Japan, Datsun copied it without a license.

Figure 78. Austin Seven, 1931. Assisted only by a single draftsman and against the convictions of his fellow directors, Herbert Austin designed a scaled-down but full-featured car capable of carrying four persons. After a slow start in 1922, Austin's "Baby" developed into one of the most successful cars in Europe and was offered in various versions until 1939. Four-cylinder engine, 750 cc, 13 hp (9.5 kW).

In 1925, Fiat began assembly line production of its small car, the 509, of which 90,000 were built. Morris was able to build more than 54,000 examples of its Cowley and Oxford models in a single year (1925), thereby topping every other English marque in sales numbers. Morris also established a far-flung network of service garages.

These successful models by Citroën, Austin, Morris, Fiat, and Opel—all with front-mounted, water-cooled four-cylinder engines displacing about one liter and driving the rear wheels— gave new meaning to the small-car concept. Their technology, although simple, shared nothing with the primitivism of the tricycles and quadricycles of the prewar years. They were considered reliable, undemanding basic transportation, and swept cycle cars (Figure 79) and minicars, which had appeared in Europe en masse after the war, from the marketplace.

Figure 79. Mollmobil, 1924. This cycle car built in Chemnitz, with readily apparent primitivism—wire cable steering, chain drive on spool (no differential) rear axle— reveals its amazing features only on close inspection. These features include an oversized central box structure, which houses the powerplant (mid-engine) and at its ends the suspensions; the driver and passenger sit inside the open box. The steel sheet- metal body serves no structural function. Hung off the body are fenders, running boards, and a right-side door. Other characteristics include a front swing axle suspension and four-wheel brakes. Approximately 1,500 were built in 1924–1925. Single-cylinder two-stroke engine by DKW, 170 cc, 3 hp (2.2 kW).

Under pressure from imports, other auto manufacturers attempted to imitate American fabrica- tion methods but often produced cars that completely missed the market. Audi maneuvered itself into financial difficulties with overly large and expensive cars (and Rickenbacker engines, the tooling for which had been bought from the American firm by Audi). The same was true of

Stoewer. In 1927, Adler introduced its "Standard 6," intended as the German answer to American imports. However, in the hands of its outside designer, Gabriel Becker of the Charlottenburg Technical University, it was no more than a Germanized American car (Figure 80). With the Standard 6, Adler introduced hydraulic brakes, all-steel bodywork, and central lubrication to Germany—all bought from German suppliers Teves, Ambi-Budd, and Vogel, the German licensees of Lockheed, Budd, and Bowen, respectively. All-steel bodywork had already been introduced to Europe by Citroën in 1925, again as a Budd licensee.

Figure 80. Adler Standard/Favorit, 1927–1934. A conventional appearance hides intrinsic virtues: the first German production car with hydraulic brakes, all-steel body, and central lubrication based on American licenses, as well as large-scale application of aluminum alloys. These were built using a modular system, producing the four-cylinder Favorit, the six- and eight-cylinder Standard with otherwise identical 2-, 2.5-, 3-, and 4-liter engines in various body styles. The example shown here is a six-cylinder 2916 cc, 50 hp (37 kW).

Since the end of the 1920s, body design, once the product of craftsmen and engineers, began to come under the influence of American industrial designers. From around 1927 to the 1950s and beyond, they determined the body styling of passenger cars throughout the world. However, they also shaped household appliances, railroad trains, cigarette lighters, furniture, and packaging for cigarettes and other consumer and capital goods.

The most famous of these designers were J. Frank de Causse (Locomobile and Franklin), Harley J. Earl (General Motors), Norman Bel Geddes (Graham, Chrysler, and Nash), and Walter Dorwin Teague (Marmon), as well as Raymond Loewy and Brooks Stevens, both of whom worked for Studebaker. The new wave of body styling was characterized by soft curves and transitions to the roof, hood, and fenders, and was soon copied in Germany. While Stoewer copied the Gardner, Horch's eight-cylinder 13/65 HP of 1928 drew heavily on Harley Earl's Cadillac derivative of 1927, the LaSalle (Figures 81 and 82).

Figure 81. LaSalle, 1927. In 1927, with an eye toward potential customers slightly below the luxury class, General Motors introduced a new marque, LaSalle, which offered the same quality and equipment as Cadillac but with a shorter wheelbase and smaller exterior dimensions. Unusual for its time was the General Motors decision to have a designer (Harley J. Earl) create its body styling. V8 engine, 4965 cc, 80 hp (59 kW).

Figure 82. Horch 8 Type 305, 1928. In 1926, Horch brought out a luxury car featuring a complex engine design but rather staid body styling. In 1928, it was given more pleasing lines, with an unmistakable similarity to the 1927 LaSalle. More than 10,000 examples of the 305 were sold, a respectable achievement for an eight-cylinder car in difficult times. Eight-cylinder engine, 3376 cc, 65 hp (48 kW).

However, long before that, German body manufacturers and graphic artists had begun to apply the styling direction proposed by American industrial designers. For the 1908 Prince Henry Trials, August Horch (1868–1951) had coachbuilders Ludwig Kathe & Sohn build an extraordinarily rakish body with door cutouts and an unbroken beltline (the "Cigar," Figure 83). Decades later, the British trade magazine *Old Motor* described the Prince Henry Horch as "...a distinct milestone in the development of coachwork...[and] the dawn of a new era in coachbuilding" (May/June 1973, p. 235–236).

Figure 83. Horch 11/22 HP, 1908. The displacement and weight requirements of the 1908–1910 Prince Henry Trials, contested with fully loaded touring cars, prompted Horch to reduce aerodynamic drag with the goal of achieving higher speeds than the competition, without increased engine output. Three doorless "cigars" were built by Kathe to his own design ideas, with continuous beltlines and separate fenders. Shown here are August Horch with his adopted son Eberhard. Four-cylinder engine, 2722 cc, 22 hp (16 kW), top speed approximately 72 km/h (45 mph).

Based on the Horch Zigarre ("Cigar"), in 1909 Kathe developed a body style for ordinary touring cars incorporating its "torpedo" shape (Figure 57, No. 2), doors, and continuous beltline. This marked the end of the baroque, French-influenced styling era exemplified by earlier cars and initiated a new, rational vehicle architecture with smooth side panels. This found its culmination in the body shapes introduced from 1911 onward by the coachbuilders firms of Kathe, Kellner (Figure 84), and Tönjes, described by graphic designer Ernst Neumann-Neander as "the architecture of curves" in the *Jahrbuch des Deutschen Werkbundes 1914* ("Yearbook of the German Werkbund," the Werkbund being an association of architects, designers, and representatives of the crafts and industry).

Figure 84. Horch 13/35 HP, 1911. Inspired by the Prince Henry Trials, several German coachbuilders created an "architecture of curves." This was marked by a cleaned-up external appearance, sometimes with rather daring lines at the front and rear of the roof, emphasized by curved windshields and frameless side glass. Seen here is a torpedo sedan by Alexis Kellner of Berlin. Four-cylinder engine, presumably 3176 cc, 35 hp (26 kW).

Before World War I, the "architecture of curves" had created "a unique and completely independent German style, a surprisingly good style which is, for the first time in matters of taste, the leader in its field." (AAZ 44/1911, p. 34) Remarkably, after the war, this attractive style, which radiated mobility, was replaced by the so-called "German body style," whose pointed radiator, wedge-shaped windshield, and boxy overall appearance imparted a static look and did not appeal to the public (Figure 85). No wonder then that curves *à l'américaine* found favor among German customers and were readily adopted. German coachbuilders could have saved themselves the detour to the United States if they had continued to develop their prewar direction.

While designers reentered the long-neglected field of sculpting ordinary and everyday items, architects, who occasionally turned their attention to the automobile, regarded their own designs as the expression of a new society and civilization, whose substance they wished to determine. For this reason, a vehicle designed by an architect usually not only appears different from ordinary cars but also often questions the shapes that have evolved over several decades. To a certain extent, this is equally true of aircraft designers who try their hand at automobiles.

Figure 85. NSU 14/40 HP, 1922. A typical representative of the post-World War I "German body style," with pointed radiator, V-shaped windshield, sharply creased rooflines, and a hard beltline. A tucked-in waistline, oval windows, round fenders, and door handles could not mitigate the stiff and cluttered overall impression. This body, with a removable sedan "pavilion" (the entire structure above the beltline), probably was built by Schebera of Heilbronn. Four-cylinder engine, 3610 cc, 40 hp (29 kW).

Examples of vehicles developed by architects include Richard Buckminster Fuller's "Zoomobile" of 1927 and his "Dymaxion" of 1932–1935 (Figure 86). Frank Lloyd Wright's "Automobile with a Cantilevered Top" of 1920 and his "Road Machine" of 1958 were unorthodox vehicles. More conventional was Le Corbusier's 1927 design for a "Maximum Car" and Walter Gropius' 1930–1931 body stylings for Adler. None of these influenced the subsequent development of the automobile or its body styling.

Nevertheless, with the exception of Gropius, these architects sought to achieve the most stream-lined shapes for their "driving machines." In the 1920s, aerodynamics played a major role, at least in the trade journals and lectures presented by forward-thinking men. Several prototypes were built; however, in the final analysis, these—similar to the automatic transmission, diesel passenger car, and all-steel bodywork—were among the developments that would not find their way into production for at least another decade.

Among those pioneers who even before 1920 dared to reduce the wind resistance of family cars were Grégoire (1912) and Castagna (1913) (Figure 87). Beginning in 1920, Austrian airship and airplane designer Paul Jaray conducted the first scientific research into automobile aerodynamics. Using streamlined bodywork, he sought to reduce the amount of dust whirled up on the unpaved roads of the time and reduce fuel consumption. Secondary goals were to increase top speed and improve ventilation. The shape developed by Jaray, combining a wing

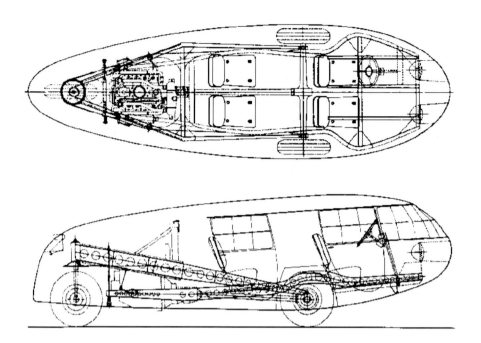

Figure 86. Fuller Dymaxion, 1933–1935. Completely ignoring well-established automotive design parameters such as good handling, vehicle dynamic stability, space utilization, compact design, outward visibility, and weight, Fuller commissioned three examples of his Dymaxion, of which one survives. The rear-mounted engine drove the live front axle; the steerable rear wheel was supported by a second frame. This three-wheeler, approximately 6 meters (20 ft) long, with a 1.8-meter (6-ft) front overhang, had wooden ribs sheathed in aluminum panels. Ford V8 engine, 3622 cc, 65 hp (48 kW).

section for the car body with a shape resembling half an airship tacked on for the greenhouse, was far from ideal. The theoretically correct approach, with laminar flow along most of the body, would have resulted in an extremely long tail. Jaray proposed a more or less steeply raked rear section with a vertical lower valance. This resulted in early flow separation, with corresponding negative effects on the drag coefficient.

Under contract to auto manufacturers Ley, Audi, Dixi, and Apollo, several coachbuilders produced prototypes (Figure 88) and race cars based on Jaray's designs. To the public, these came across as inorganic and futuristic. Only with the introduction of a lowered chassis in the 1930s did Jaray's ideas regain momentum, even if the aerodynamic qualities of later bodywork usually suffered from protruding fenders, headlights, and radiators.

A year earlier, 1921, aircraft maker Edmund Rumpler had exhibited an unconventional car at the Berlin auto show. If there were a sensation at that show, it was the *Tropfen-Auto*—the Teardrop Car (Figure 89). This, at least, is the impression imparted by a veritable storm of enthusiastic reports in the domestic press.

Figure 87. ALFA 40/60 HP Castagna, 1913. Commissioned by Count Ricotti, Carrozzeria Castagna of Milan built a streamlined body atop a conventional ALFA chassis (later known as Alfa Romeo). The body is strongly reminiscent of a 1910–1911 design by body engineer Oskar Bergmann, who had suggested wheels tucked into the bodywork, more conventional windows instead of portholes, two doors instead of one, and an underside fairing. A replica ALFA was built in 1979. Four-cylinder engine, 6079 cc, 60 hp (44 kW), sometimes also listed as 82 hp (60 kW).

Figure 88. Jaray prototypes, 1922–1923. To advance the concept of streamlining, Jaray organized promotional tours throughout Germany. Although top speeds were 20–25 km/h (12–16 mph) higher and fuel consumption about 20 percent lower than contemporary production models, the public rejected these futuristic cars. From left to right: 1922 Ley T6 with body by Spohn, 1923 Audi K 14/50 HP, and 1923 Dixi 6/24 HP, the latter two with bodywork by Gläser.

Figure 89. Figure 89. Rumpler Tropfen-Auto 10/30 HP, 1921. Not to be outdone by architects, aircraft makers (Dorner, Dornier, Fend, Grade, and Rumpler) built unorthodox cars of their own. With the exception of the 1953–1964 Messerschmitt microcars developed by Fritz Fend, production numbers remained quite small. This included production of the Rumpler Tropfen-Auto (Teardrop Car). Between 1921 and 1925, fewer than 100 examples were built. Six-cylinder engine in W configuration, 2580 (or 2308) cc, 35 hp (26 kW).

Indeed, with his Teardrop Car, Rumpler had implemented ideas that were far in the future and, to some extent, have not been realized to this day. Above all, his streamlined bodywork attracted considerable attention. Only wheels and suspension pickup points extend beyond the body profile. Shift and handbrake levers, headlights, and horn, normally mounted outside the bodywork, are enclosed. Horizontal fins replace sweeping fenders; the front-hinged doors and curved glass became commonplace only in the 1950s. Full underbody fairings and central steering have not yet been realized. The drag coefficient (c_d) of 0.28, determined in the Volkswagen wind tunnel in 1979, set a record unmatched for decades (Table 5).

To reduce the unsprung mass of conventional live rear axles and improve ride comfort, Rumpler employed his own patented swing axle. Although it did not perform with complete satisfaction in the Teardrop Car, it nevertheless represented a vital improvement in passenger car design. After it was tried by Tatra in 1923 (Figure 90) and Steyr in 1926, it found general acceptance in French and German passenger car designs of the 1930s, giving these a more comfortable ride than contemporary British, American, or Italian cars.

TABLE 5
DRAG COEFFICIENTS (c_d)

		c_d	Figure No.
1921	Rumpler Tropfenwagen	0.28	89
1927	Adler Standard 6	ca. 0.70	80
1937	DKW Meisterklasse	0.585	102
1934	Tatra 77	ca. 0.37	107
1953	Borgward Hansa 2400 Fastback	0.43	108
1966	Volkswagen Type 1 (Beetle)	ca. 0.48	151
1967	NSU Ro 80	0.355	163
1974	Volkswagen Golf (Rabbit)	0.42	165
1980	Citroën GSA	0.34	—
1981	Audi 100	0.30	—
1987	Opel Omega	0.28	—
2000	Audi A 2 1.2 TDI 3 L	0.25	213

NOTE: Drag coefficient is a theoretical number measuring air resistance. It may be altered by the shape and surface condition of a body, and can be calculated or determined experimentally in a wind tunnel. Measurements may differ because neither wind tunnels nor test methods have been standardized. Comparison of c_d for different vehicles is therefore problematical.

Figure 90. Tatra 11 suspension, 1923. Rumpler's work with swing axles derived from his days with Adler (1902–1905), where he experimented with swinging halfshafts and a drivetrain mounted in, rather than on, the frame. Although his first efforts were unsuccessful, Rumpler continued to patent ever-improving swing axle designs—possibly encouraging Ledwinka to equip his Tatra designs with rear swing axles. Of note are the tubular backbone chassis and air-cooled boxer engine. Two-cylinder horizontally opposed engine, 1056 cc, 12 hp (9 kW).

Also unconventional was its engine location, ahead of the rear axle (Figure 91), a move forced by the shape of Rumpler's chassis and bodywork. Advantages of this layout include compact design and elimination of the driveshaft; drawbacks include limited accessibility and increased interior noise. Incorrectly described as a mid-engine layout, it would one day become standard practice in race car design. As early as 1923, Benz acquired a Rumpler license to build its Benz *Tropfen-Rennwagen* (Teardrop Racer). Auto Union followed in 1934, and English racers followed after World War II. Since the 1960s, there have been virtually no purebred race cars whose engines have not been located ahead of the rear axle.

Despite encouraging development work, until around 1930 most European passenger cars followed American design trends or, more accurately, the standard layout presaged by Maybach in 1900–1901: an upright four- or six-cylinder front-mounted engine, rear-wheel drive and beam, and live axles front and rear (Figure 74). Nevertheless, left-hand drive, standardized pedal arrangement, and four-wheel brakes, usually mechanically actuated, had become the norm. Enclosed, box-shaped bodywork had replaced the open touring car, but without any regard for aerodynamics (Figure 73).

Figure 91. Rumpler Tropfen-Auto chassis, 1921. With its chassis layout dictated by the teardrop bodywork it carried, the Rumpler Teardrop Car displayed a break with traditional ladder frame practice. Instead, it has a deep frame (with a double floor, similar to the 1997 Mercedes A-Class) with a well for two spare tires, openings for axles, springs, and linkages. The exhaust muffler is in the rear apex of the frame; the console supporting the steering box is in the front. The engine, mounted ahead of the rear axle, was initially one of "W" cylinder layout and later an inline four-cylinder.

The Automobile Industry and Automotive Technology in the 1930s

In the years before World War I, the automobile played no significant economic or political role. Seen from an economic standpoint, the mostly small firms did not employ many workers, and the infrastructure was underdeveloped. Railroads were regarded as the dominant means of land transportation. In the 1920s, the automobile matured into an economic factor, a status that it had long held in the United States. In the 1930s, the automobile and its development were closely tied to the fate of individual nations.

The United States, as an up-and-coming world power, had established assembly plants in many countries or had bought firms that produced models suitable for the home markets. In production and marketing, less so in engineering, the United States set the standard. In 1932, one in five Americans owned a car; in Germany, only one in a hundred. Regardless, the American share of worldwide passenger car production dropped from 85 percent (4.5 million units) in 1929 to slightly less than 66 percent (2 million units) in 1938. Germany's share in 1937 amounted to only 5 percent.

As a world power in decline, Great Britain was happy in its own "splendid isolation." Automobiles designed for the home market, with its narrow and winding roads, had little appeal in terms of style or technology to buyers in continental Europe or in open (i.e., non-Commonwealth) overseas markets. Because of well-protected home markets and its colonies, the United Kingdom managed to retain its second place in worldwide car production behind the United States (Table 6).

TABLE 6
WORLDWIDE PASSENGER CAR PRODUCTION
1933 AND 1938

	1933		1938	
	Units	%	Units	%
USA	1,573,512	72.9	2,000,985	65.6
England	220,775	10.2	341,028	11.2
Germany[1]	92,226	4.3	276,804	9.1
France	163,770	7.6	199,750	6.6
Canada	53,849	2.5	125,081	4.1
Italy	32,000	1.5	59,000	1.9
USSR	10,252	0.5	26,800	0.9
Czechoslovakia	8,670	0.4	12,500	0.4
Japan	191	–	8,500	0.3
Austria	1,150	0.05	–	–
Belgium	800	0.04	1,075[2]	–
Others	980	0.05	N.A.	–
TOTAL	2,158,175	100	3,041,948[3]	100

RDA 1937 & 1938. Figures in some cases approximate.
1. Includes Austria as of July 1938.
2. 1939.
3. Excluding Belgium and "others."

France, once a leading car-building nation, suffered under social tension and economic and financial crises. With its antiquated production facilities, it dropped behind Germany in new car production in 1935. Paralyzed, French car makers observed the drive to innovate and the aggressive market policies of their countryman André Citroën (1878–1935).

Italy, weakened by the economic crisis that began in 1927, was forced to export most of its automobile production to buy coal, oil, iron, and other raw materials. The domestic market remained underdeveloped.

Because of the progressive Tatra models designed by Hans Ledwinka, the once insignificant Czech auto industry found itself catapulted into the vanguard of automotive technology, to the point where it sold manufacturing licenses. Meanwhile, the automotive landscape of Belgium evolved in the opposite direction. By 1939, the once-great marques Metallurgique, Minerva (Figure 92), Excelsior, Imperia, Nagant, and FN had disappeared, because of ill-considered customs policy and product planning. Instead, outside investors, primarily American, established more than ten assembly plants in Belgium. Antwerp served as the European distribution center for Detroit.

Figure 92. Minerva 32 CV AKS, circa 1929. From 1909 to 1935, Minerva used only the quiet-running Knight sleeve valve engines (Figure 51). Shown here is a 2+1 coupe built around 1935, bodied by an unknown coachbuilder, with skylights above the windshield, double rear cabin wall, and oversized boxes alongside the hood for spare tires and luggage. Six-cylinder engine, 5950 cc, 150 hp (110 kW).

Germany prospered under the auto-friendly Nazi regime. Hitler recognized the automobile as a means to achieving his ends and supported the auto industry by means of tax incentives, road and autobahn construction, state subsidies for race car design and construction, motorsports events, and the elimination of the Reich railway system's control of road transport tariffs.

At the end of the 1920s, there was already a hint of the dichotomy that would become apparent to everyone in the 1930s: the American preference for large cars and, due to comparative economic weakness, the European penchant for small or mid-sized cars. This gap would not be closed until the energy crises of the 1970s.

In North America, automobile development led to large, thirsty engines with relatively low power output but high torque. In Europe, the trend was to small, high-revving, economical engines. Even Ford, who claimed to build affordable cars for the common man, shifted to a mass-produced car powered by a V8 engine (Figure 93). European Ford customers, on the other hand, had to be satisfied with a four-cylinder car, developed in England as the Model Y, which was also built and sold in Germany as the Köln (Cologne) (Figure 94).

Figure 93. Ford V8, 1932. Chevrolet's successful International 6, with its six-cylinder engine, proved a difficult competitor for the Ford Model A, successor to the Model T. In response, Ford introduced a V8 model in 1932, attractively priced at slightly less than $500. By 1935, a million examples had been built, of an eventual 15 million. Production facilities included a plant in Germany. The V8 engine was built in the United States until 1954 and in Germany until 1961 (commercial vehicles). V8 engine, 3622 cc, 65 hp (48 kW).

Figure 94. 1934 Ford Model Y/1937 Ford Eifel. Ford of England, originally based in Manchester, began operations in 1911 by converting American left-hand drive models to right-hand drive. Ford opened its new Dagenham plant in 1932 with the Model Y, its first all-British design. In 1933, this small one-liter car was adopted by Ford's German plant and marketed as the Köln (Cologne). It was succeeded by the Eifel (production life 1935–1939), itself a German version of the British Model C (also 1935–1939). Model Y 8 HP/Köln: four-cylinder engine, 933 cc, 21 hp (15 kW); Model C 10 HP/Eifel: four-cylinder engine, 1172 cc, 34 hp (25 kW).

From 1930 onward, American automotive technology began to lose its pre-eminence over that of the Old World. While European manufacturers had spent the 1920s mainly in developing engines for increased power and reliability, the 1930s saw chassis and bodywork taking the lion's share of attention. The Old and New Worlds developed into equal partners; the advancements achieved by one also benefited the other.

One example is the automatic transmission. Even before World War I, Hermann Föttinger and his colleague Wilhelm Spannhake had been granted patents on fluid clutches and transmissions. Working at Berlin's Forschungs- und Prüfungsanstalt für Windkraftmaschinen (Wind Power Research and Test Establishment), founded in 1925, Föttinger developed a hydraulic transmission (Figure 95). In late 1936 and early 1937, he conducted successful test drives with this installed in a Mercedes. This pioneering German effort was followed by General Motors' push to develop the concept for mass production, and the General Motors Oldsmobile division first offered its "Hydramatic" transmission in 1940.

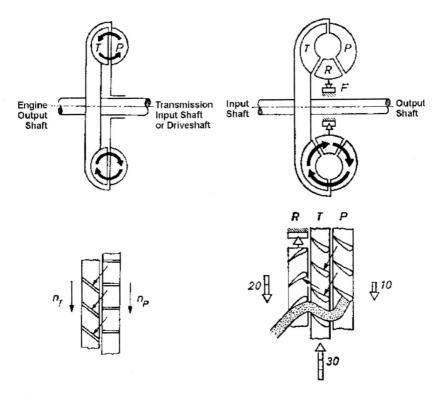

Figure 95. Hydraulic coupling (left) and torque converter (right) after Föttinger (schematic drawing by Dr. Ing. J. Loomann, Friedrichshafen). Coupling: with the engine rotating, hydraulic fluid from the impeller P impinges on the turbine T, exerting pressure on the turbine blades. This produces a torque. Automatic transmission or torque converter: an additional reactor R diverts the stream of hydraulic fluid from T, resulting in a torque increase of T relative to P. F = free wheel. Because torque converters operate economically only at medium speeds, they are combined with geared transmissions.

A second example is the "silent" manual transmission. As early as 1900, the Diamant Speed Gear Company of Trieste (then part of Austria) had built transmissions with constant-mesh helical gears. Compared to straight-cut gears, these were easier to shift, produced less noise, and showed lower wear rates because the increased length of contact along the tooth faces distributed loads over a greater area. It took more than 20 years before American transmission manufacturers and auto makers employed helical-cut constant-mesh gearboxes, but the Europeans took even longer to adopt the concept. Zahnradfabrik Friedrichshafen (ZF) first offered its three-speed "Aphon" transmission, with helical-cut gearsets, in 1929.

A third example is front suspension. In 1933, Mercedes introduced independent front suspension via dual transverse arms and coil springs. General Motors followed a year later with two completely different independent front suspension systems, both called "Knee Action." One system was for Buick, Oldsmobile, and Cadillac, a transverse short and long arm suspension with coil springs and lever-action shocks. The other, a short-lived design for Chevrolet and Pontiac, was a modification of Dubonnet's trailing arm "knee action" principle with integral spring/shock units. Other manufacturers on both sides of the Atlantic soon followed with independent front suspension of their own.

Italian, British, or American customers spent little time pondering the merits of two- or four-stroke engine, front- or rear-wheel drive, live or independent axles. (However, the German mentality being what it was, these were grounds for religious warfare.) For the time being, they stuck to their "skyscrapers on wheels, with their deadly high centers of gravity" (*Motor-Kritik*, Vol. 12/1931, p. 261). More properly, they held onto the standard layout with its beam and live axles and tall frame. By 1935, many cars featured double cranked frames with transverse leaf springs. Some models employed swing axles suspended by coil springs and located by trailing arms. This European technological superiority also manifested itself in higher export figures. The so-called "American Peril" was no longer a concern.

Independent suspension improved ride comfort, increased active safety, and permitted lowered bodywork and lower step-in height. These new chassis were prerequisites for more streamlined bodywork. The freestanding radiator disappeared behind a grille that, similar to the windshield, was raked. Instead of a luggage rack and luggage exposed to the elements, bodies were given an enclosed luggage compartment—a trunk. By the end of the 1930s, this had been made accessible from the outside of the car and was given a streamlined shape (Figure 96).

Improvements to auto body shapes soon were matched by new manufacturing technology. General Motors figured that when produced in numbers greater than 40,000, a unit body (combining both chassis and bodywork into a self-supporting, inseparable structure) would not only save weight but also reduce costs. Developed by Theodore Ulrich and originally intended for a small Chevrolet, the sheet-metal structure was introduced by General Motors' European subsidiaries Opel (1935, Figure 97) and Vauxhall (1937). The underlying reason was that with its frequent facelifts and numerous body variations, this inflexible system did not fit into the American manufacturing landscape. The American auto industry did not convert from body-on-frame to unit bodywork until 1959–1960. European plants had made the switch five to ten years earlier.

Obviously, unit bodywork did not strike the world of automotive technology as a bolt from the blue. It also had its antecedents. For example, all horse-drawn carriages that did not have a chassis (running boards, later perches) as rigid intermediaries between suspension and coachwork actually had unit bodies, which absorbed the various suspension forces. In 1906, the French firm Alcyon and its German licensee Hansa employed a combined frame and lower body section as a structural element of their voiturettes. In 1921, aviation pioneer Hans Grade designed a single-seat cycle car with a self-supporting sheet-metal shell. Wunibald Kamm

Figure 96. Hansa/Borgward 2000/2300/3500, 1937–1942. The Borgward works were always open to new ideas, as shown here by their prewar six-cylinder class: a "streamlined" four-door sedan or convertible sedan, both without running boards, a choice of various engines (modular system), central box frame, and front and rear swing axles, as well as hydraulic brakes. Six-cylinder engine, 1962/2247/3485 cc, 48/55/90 hp (35/40/66 kW).

Figure 97. Opel Olympia 1.3-liter, 1935–1937. The first volume-production automobile with a unit body, consisting only of sheet steel structural members and stampings, with the roof and roof pillars also sharing structural loads. To familiarize the public with this unfamiliar concept, Opel displayed a sectioned Olympia at the 1937 Berlin International Automobile and Motorcycle Exposition (IAMA). The vehicle has been lost, but in 1977 a replica was built (shown here), at present on display in the Deutsches Museum in Munich. Four-cylinder engine, 1288 cc, 26 hp (19 kW).

composed the four-seat SHW of 1924–1925 with an aluminum alloy unit body. With its Lambda of 1925 (Figure 98), Lancia replaced the customary frame longitudinals with a pair of deep-section sheet steel members, joined to each other by the dashboard, seat consoles, and rear transom. The Citroën Traction Avant (1934–1957, Figure 99) represented a transition stage between frame-on-chassis and unit body construction. It consisted of a basic cage to which various subassemblies were bolted. The basic cage, composed of engine mounts, firewall, and transverse and longitudinal members, derived from a patent by Edward G. Budd Manufacturing Company of Philadelphia.

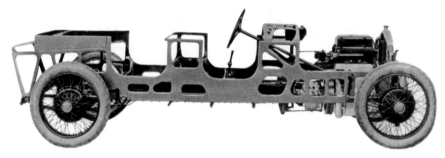

Figure 98. Lancia Lambda, 1923. Vincenzo Lancia replaced the longitudinal frame members of the conventional ladder frame with deep stamped steel sections, which gained their stiffness from transverse members such as the dashboard, seat boxes, and transom. The front and rear suspension members are carried by fabricated steel structures (partially hidden). The Lancia's riveted frame and lower bodywork structure may be regarded as a semi-unit body. V4 engine, 2120 cc, 49 hp (35 kW).

Figure 99. Citroën 7 CV Traction Avant, 1934–1957. The semi-unit body of the Citroën consists of a basic frame of stamped steel members, encompassing rocker panels and firewall, with pickup points for the rear axle and front powertrain assembly. The transmission is mounted ahead of, and the engine behind, the front axle. Front-wheel drive and a long wheelbase gave above-average road manners. Four-cylinder engine, 1911 cc, 56 hp (41 kW).

Generally, this Citroën 7 CV incorporated the advanced thinking of the time with realizable progress similar to no other mass-produced car: long wheelbase, front-wheel drive, "Floating Power" engine mounts, independent front suspension, trailing arms at the rear, front and rear torsion bars, hydraulic brakes, later rack-and-pinion steering, and telescoping shock absorbers. This exemplary design by André Lefèbvre manifested itself in outstanding handling and a long production life.

Wealthy customers continued to order custom-built coachwork on luxury chassis. No shortage of such business existed in the 1930s. Packard, Franklin, Lincoln, and Pierce-Arrow in the United States, British Daimler and Rolls-Royce, and on the Continent Horch, Maybach, and Voisin offered complete cars or chassis with 12-cylinder engines. Cadillac, Marmon, and the small French specialty maker Bucciali even built cars with 16-cylinder powerplants. Against these, 8-cylinder cars seemed almost unworthy of mention, although Duesenberg in the United States and Bugatti in France (Figure 100) built truly majestic automobiles. Not surprisingly, the 1930s are often considered the classic era of the automobile.

Figure 100. Bugatti Type 41 Royale, 1932. A classic par excellence: Ettore Bugatti's ultimate achievement, with truly gargantuan proportions. The wheelbase alone was 4.3 meters (14 ft); the engine is 1.4 meters (4.6 ft) long. Even kings and potentates, for whom the Royale was intended, shrank from the prospect of buying one. Only six were sold, each with custom coachwork. Shown here is a cabriolet by Ludwig Weinberger of Munich, at present in the Henry Ford Museum in Dearborn, Michigan. Eight-cylinder engine, 12,763 cc, 300 hp (221 kW).

The Renaissance of German Automotive Technology

The preceding years had shown that automobiles built to American tastes were too large and too expensive to maintain, and therefore only marginally suitable for European conditions. After 1931, the wealth of ideas demonstrated by one-off products or limited production runs in the 1920s precipitated new design solutions for mass-produced vehicles, and the German auto industry assumed a leading role. It also introduced two-strokes and diesel engines as specifically German contributions to automotive engineering.

Apart from more or less successful front-drive outsiders that had come and gone since 1897, including the English Alvis of 1928, the American Cord of 1929, and various French cars around 1930 employing J.A. Gregoire's front-drive concept, DKW and Stoewer broke away from the "standard layout" and in 1931 introduced small cars with front-wheel drive. They were followed a year later by Adler, and in 1933 by Audi and NAG. In France, Citroën in 1934 (Figure 99) and Amilcar in 1937 followed their lead. Most of these had an axle-less (i.e., independent) front suspension by transverse springs and transverse arms, combined with a light rear axle, hydraulic brakes, rack-and-pinion steering, and lowered frame made possible by elimination of the driveshaft. These advanced front-drivers offered more active safety than more expensive cars with live axles and rear-wheel drive.

To make ordinary "standard layout" cars safer and more comfortable, Daimler-Benz introduced its 1931 Type 170 with a rear swing axle on coil springs. In 1933, it presented the Type 380 with the aforementioned front suspension via transverse links and coil springs. Both features were quickly adopted by other manufacturers.

Hopes that the front-drive and swing axle models of 1931 would resuscitate the overall German automotive economy were not realized because of the continuing world economic crisis. Quite the opposite occurred. Passenger car production dropped from 77,333 units in 1930 to 62,563 in 1931, and further to 43,448 units in 1932, including assembly plants (assembling cars shipped in CKD form—completely knocked down). In those years, the only gains were marked by exports and three-wheeled vehicles *à la* Goliath (Figure 101). Because of long-standing regulations, these did not require driver's licenses and were tax exempt if they did not exceed 350 kg (772 lb) empty weight and 200 cc displacement.

To help "that perennial whipping boy, the motor vehicle" (*ATZ*, Vol. 17, 1932, p. 401) back onto its feet, various associations, motoring publications, and private individuals repeatedly pointed out the enormous growth in the automotive sector's impact on the economy since World War I and offered proposals for its revival. They recommended reductions in general or vehicle taxes, fuel import duties, and insurance premiums; replacing the motor vehicle tax with a fuel tax; building roads and autobahnen to reduce unemployment; elimination of monetary controls; removing the conflict between the railroad and automotive sectors; offering a "people's car" for 800 Marks; and finally elimination of the fuel-blending requirement.

Figure 101. Goliath Pionier, 1931–1934. Despite the global economic crisis, in the 1930s cycle cars experienced unusual popularity in Germany, traceable to fact that they were tax free and that their operation did not require a driver's license. The most successful model was the three-wheeled Goliath Pionier, of which approximately 4,000 were sold. Choices included the "world's cheapest sedan" (as claimed on the back of this postcard), a convertible, a phaeton, a sport two-seater, and even a "streamlined sedan." Single-cylinder ILO engine at the rear, 198 cc, 5.5 hp (4 kW).

After the National Socialists seized power on January 30, 1933, Hitler also seized on some of these ideas and put them into motion. Because the 1931 motor vehicle laws were due to expire on April 1, 1933 in any case, Hitler was able to introduce motor vehicle tax relief almost immediately. All passenger cars and motorcycles registered after April 1 of that year were completely exempt of taxes.

In Germany, the economic recovery, which had taken hold worldwide by the end of 1932, accelerated rapidly thanks to further actions by Hitler's auto-friendly administration. The demand for motor vehicles escalated, and new plant hires were the order of the day. By 1937, Germany enjoyed full employment. The world economic crisis that had peaked in 1932 was finally a thing of the past.

Just before the end of 1931, the State Bank of Saxony pressured the independent firms Horch, Audi, the car division of the Wanderer Werke, and the Zschopauer Motorenwerke (DKW) to merge under DKW leadership, with headquarters in Zschopau (relocated to Chemnitz as of

1936). The success of the DKW two-stroke motorcycles had prompted founder Jörgen Skafte Rasmussen to build two-stroke cars as well. Before and after World War II, DKW two-strokes were as much a part of the German street scene as the Ford Model T in the United States.

Following the rear-drive two-stroke cars, built from 1928 onward, the front-drive products introduced in 1931 caused quite a stir because of their attractive pricing and unconventional technology. The two-stroke two-cylinder engine was mounted transversely behind the front axle and drove the front wheels. Suspension was by transverse springs front and rear, and the plywood body was covered with leatherette. Beginning in 1935, the later Meisterklasse and Reichsklasse compact cars, with their attractive bodywork, proved to be durable and economical transportation (Figure 102). Through 1940, approximately 230,000 examples were built.

Figure 102. DKW Meisterklasse F7-700, 1937. For the 1937 IAMA, DKW sectioned one of its "Meisterklasse" models down the centerline to show off its engineering features and interior space. Typical for DKW were its two-stroke engine mounted transversely behind the front axle, front-wheel drive, and a wood-framed body with plywood sheathing covered in leatherette. Unusual by modern standards is the fuel tank location between the engine and dashboard, with a pass-through for the shift linkage. Two-cylinder two-stroke engine, 687 cc, 20 hp (15 kW).

The DKW two-stroke engine derived from a design by Hugo Ruppe, which DKW employed as of 1919 to power generators, toys, bicycles, and, from 1921, motorcycles and "motor chaises," the ancestors of motor scooters. Because the engine layout—with its three-port system, crankcase compression, and baffled pistons—was only marginally suited for cars weighing more than 530 kg (1168 lb), in 1929 DKW temporarily switched to a two-stroke V4 engine, with two scavenge pumps. A satisfactory relationship between power, consumption, and smoothness was not achieved until 1933, with the introduction of the three-port system employing loop scavenging based on Adolf Schnürle's patents (Figure 103). DKW engine technology dominated the two-stroke vehicle sector of the 1930s and served as a starting point for further European developments after World War II.

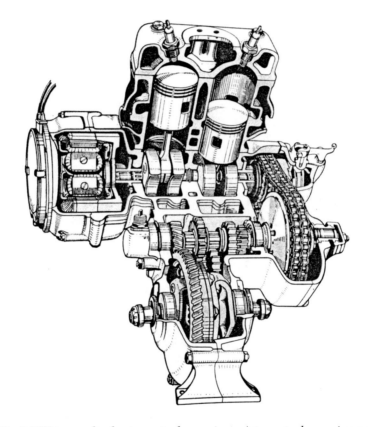

Figure 103. DKW two-cylinder two-stroke engine. A two-stroke engine produces one power stroke every other stroke, or once per revolution. A four-stroke produces one power stroke followed by three non-working strokes, or one every two revolutions. However, a two-stroke does not provide twice as much power as a four-stroke. Because of incomplete cylinder filling, its advantage is only 30 percent. Shown here is a two-stroke built to Schnürle principles with flat-topped, ported pistons. The intake ports in the cylinder walls are at the same height as, and on either side of, the exhaust ports. As a result, intake gas flow nearly returns to its entry point—hence, the term "loop scavenging."

Four-stroke engines also enjoyed continued development. In 1938, to make its engines "autobahn rated" while keeping wear and breakdowns to a minimum, Opel introduced short-stroke engines with correspondingly lower piston speeds (Figure 104(b)). By this time, the changeover from inefficient side-valve (SV) engines (Figure 104(a)) to overhead valve (OHV) engines was complete. Scientific research such as that conducted by Sir Harry Ralph Ricardo (Comet Turbulent Cylinder Head, 1919) had shown that the large, convoluted combustion chambers of SV engines did not promote optimal combustion of the air/fuel mixture.

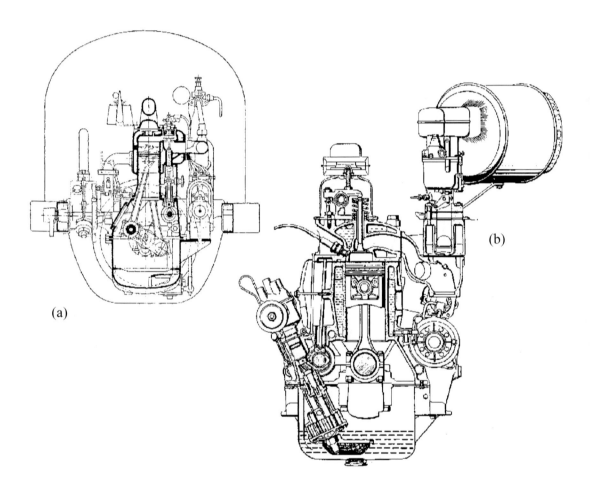

Figure 104. Engine development. (a) A typical example of an SV (side valve) engine. The combustion chamber extends to the side to encompass vertical valves; above these is a pressure release valve and/or spark plug. Bore/stroke ratio 0.62 (a so-called "undersquare" engine), compression 4.5:1. 1914–1921 Benz 8/20, four cylinders, 2093 cc, 20 hp at 1900 rpm. (b) "Autobahn-proof" OHV (overhead valve) engine. Combustion chamber with spark plug and vertical or slightly inclined valves directly above the piston, bore/stroke ratio 1.08 ("oversquare" engine), compression ratio 6.0:1. 1938–1940 Opel Olympia, four cylinders, 1488 cc, 37 hp at 3500 rpm.

The road system, originally created to meet the needs of horse-drawn vehicles, was unsuitable, both in its construction and its geographic layout, for the increases in automobile traffic during the 1920s. Particular problems included dust whirled up by traffic and potholes formed in the waterbound macadam roads that made up 90 percent of the German rural road network.

By the mid-1930s, surface dressings, asphalt emulsions, paving blocks, and concrete roads, along with improved surveying and routing of roads, allowed the road system to satisfy the needs of automobile traffic. What remained were hair-raising passages through towns and horrifying grades that made the jobs of long-distance truck drivers particularly challenging. As a result, on rural roads, average speeds for trailer-semitrailer combinations with two trailers were limited to 25 or 35 km/h (15 or 22 mph). Passenger cars, depending on engine power, weather, and road conditions, could manage 50 to 75 km/h (31 to 47 mph).

The qualitative culmination of German road building in this period were the autobahnen—expansive divided highways, free of intersections, restricted to motor traffic and intended for long-distance high-speed travel. As early as 1923, Italy, under engineer Piero Puricelli, had begun to connect Varese, Como, Bergamo, and Milan by high-speed roads. Stimulated by Puricelli's road-building plans to join all of Europe, an organization named HAFRABA was formed in Germany in 1926—a "private organization to prepare to join Hamburg (the Hanse cities), Frankfurt and Basel," with a later extension via Milan to Genoa. The first German section of roadway restricted to motorized traffic, joining Cologne and Bonn, was opened to traffic on August 6, 1932.

After Hitler seized power on January 30, 1933, the Nazis claimed both the autobahn idea and its implementation for themselves—although in 1930, they had rejected the building plans in the German Reichstag. "Adolf Hitler's roads," as they were pompously called by Reich press chief Otto Dietrich at the opening of the first Reichsautobahn between Frankfurt and Darmstadt on May 19, 1935, were "born of a National Socialist spirit, carried by the rhythm and ethics of National Socialist work; they are the concrete embodiment of National Socialist life and sense of space!"

Doubtless the "Law for establishment of the Reichsautobahn project," created in June 1933, accelerated autobahn construction in Germany. It was easily integrated into the Nazi job creation program and had strategic value for Hitler. Of the planned 7000-km (4350-mile) long network, 3065 km (1905 miles) were finished by 1938, including the major north-south and east-west corridors.

Regardless, the project was subordinate to the Reich railway system. With the Reich government as supervisory organ, it was hoped that the rivalry between rail and road, which had existed since the 1920s, would finally be laid to rest. In reality, the construction of the autobahn network laid the foundation for renewed postwar discord between the two major land transportation systems.

Comparison runs organized by the Inspector General of German Roads, Fritz Todt, proved the enormous "cultural and economic value of the Reichsautobahn, in every aspect—speed, operating costs, vehicle wear, driver workload, road safety, and driving enjoyment" (from the *Strasse* booklet series, Berlin, 1938, Vol. 10, p. 7). Indeed, to cover the distance from Bruchsal to Bad Nauheim, a 3.2-liter car covered 147 km (91 miles) on the autobahn, but 161 km (100 miles) on secondary roads; average speeds were 129 and 74.5 km/h (80 and 46 mph), respectively; fuel consumption 15.6 and 27 liters/100 km (15.1 and 7 mpg), respectively; travel time 1.14 and 2.16 hours; four gear changes compared to 102; 7.14 vs. 494.6 mkg of clutch and brake pedal work per 100 km. A further diagram in the booklet, claiming 218 oncoming cars per 100 km (62 miles) on the secondary road but not a single one on the autobahn, is good for a chuckle in view of our modern conditions.

Aside from these statistics, remember that for the first time since the Roman Empire, roads were again "faster" than the vehicles using them. Customers interpreted the top speed claims issued by auto makers as "sustained autobahn speeds," with the predictable result that they soon fell by the wayside with worn bearings, broken springs, burst tires, and damage to shock absorbers and suspension components. In response, auto manufacturers, supplier industries, and the academic community embarked on development projects. Suspension improvements, redesigned engines (short-stroke engines), and tire and lubricating oil research gave German cars the cachet of being "autobahn proof" and therefore technically ahead of foreign competitors. Even today, this argument is happily applied whenever the specter of autobahn speed limits raises its head.

The high cruising and absolute top speeds possible on "Adolf Hitler's roads" again moved aerodynamics to the forefront. This field had lain fallow since Jaray's experiments in the early 1920s.

True, in 1923 Aurel Percu of Rumania, Emile Claveau of France (1927), Sir Denniston Burney of Britain (1928), and the American firm Reo (1933) had built one-off designs which, although better streamlined, were well removed from mainstream practice. Streamlined German production cars appeared in the form of the 2.5-liter Adler of 1937 (Figure 105) and the 1.1-liter Hanomag of 1939. These achieved drag coefficients c_d of 0.36 and approximately 0.41, a considerable improvement compared to typical values of 0.56 to 0.60 for conventional bodywork of the time.

However, these German cars were not the first streamlined cars to see production. Autobahnen and aerodynamic bodywork are not inexorably tied to one another. As early as 1934, Chrysler and De Soto had premiered their sister models of the Airflow (Figure 106) at the New York Auto Show. Unfortunately, these models did not sell well, and a year later Chrysler had to introduce a redesigned model more in keeping with public taste. Europe's first streamlined production car of the new low-chassis generation was the Czech Tatra 77 of 1934 (Figure 107), designed by Hans Ledwinka. It was followed by the Singer Airstream from England, the Peugeot 402, and the Volvo PV 36, all built in 1935, and a year later the Steyr 50 designed by Karl Jenschke, who also penned the Adler 2.5-liter (*cf.* Figure 105).

Figure 105. Adler 2.5-liter Type 10, 1937. After a series of Jaray record cars and prototypes (began from 1935 onward), Adler, inexorably linked to advanced aerodynamics, brought out a full-featured five-seater in 1937 with streamlined bodywork ($c_d = 0.36$). Thanks to its forward engine location, overall height of 171 cm (67.3 in.), and lowered floorpan, its creator, Karl Jenschke, was able to provide ample interior room, thereby demonstrating the "habitability of the streamline." Six-cylinder engine, 2494 cc, 58 hp (43 kW).

Typical of Jaray's style, the bodywork of these passenger cars exhibited steeply raked rooflines. To minimize the turbulence created by these shapes, Baron Reinhard Koenig-Fachsenfeld developed a body whose tail was "cut off," forming a flat rear surface. In 1936, further research by the FKFS—the Forschungsinstitut für Kraftfahrwesen und Fahrzeugmotoren Stuttgart (Stuttgart Research Institute for Motor Vehicles and Vehicle Engines), under the direction of Wunibald Kamm, proved that vehicles with the so-called K- or Kamm tail, following Koenig-Fachsenfeld's lead, offered a good compromise between everyday utility (e.g., vehicle length and interior dimensions) and an attractive drag coefficient. In addition to aerodynamics, Kamm also placed great emphasis on vehicle stability. The first K prototype (Figure 109) was on the road in 1938. The first (and, for the time being, the last) production cars employing this principle were the Nash Airflyte (United States, 1949–1951) and in Europe the Borgward Hansa 2400 S (1952–1955, Figure 108).

Figure 106. Chrysler Airflow, 1934. Only ten years after founding his company, Walter P. Chrysler surprised the automotive world with the "first real motorcar since the invention of the automobile" (according to the sales brochure). The car in question was the Airflow, a sedan with pseudo-aero styling reminiscent of Count Zeppelin's airships and Franz Kruckenberg's "Rail Zeppelin" of 1931, and, of course, American streamlined trains. The Airflow may have inspired Jenschke's "Autobahnadler" (Figure 105). Eight-cylinder engine, 4893 cc, 122 hp (90 kW).

Figure 107. Tatra 77, 1934. The Tatra 77 (1934–1936), designed by Hans Ledwinka, set new standards with its low drag coefficient (estimated $c_d = 0.37$). Its successor, the 87, in turn improved on this, with a measured c_d of 0.36. Its streamlined shape, engineering layout with central tube chassis, air-cooled engine at the rear, subframe, and rear swing axles all significantly influenced central European automotive engineering. V8 engine, 2986 cc, 75 hp (55 kW).

Figure 108. Borgward Hansa 2400 S, 1953. The Hansa 2400 body shape, with its K-tail ("Kamm back") embodied the prewar vision of the car of the future. Although wheel cutouts, bumpers, and grille raised the c_d to 0.43, well above that of the FKFS study (Figure 109), the front contours, accentuated belt line, and raked pillars also yielded improved aesthetics. Six-cylinder engine, 2337 cc, 82 hp (60 kW), maximum speed 150 km/h (93 mph).

Figure 109. FKFS Study K 1, 1938. With its Kamm back, pontoon body, wheel skirts, smooth undertray, and curved windshield, the FKFS, under Wunibald Kamm, was able to achieve a low c_d of only 0.23, combined with good stability; however, its aesthetics left much to be desired. The coachwork was formed by Vetter; the car has been lost. Six-cylinder BMW, 3485 cc, 90 hp (66 kW), maximum speed 172 km/h (107 mph).

Most German manufacturers limited their aerodynamic efforts to increasingly rounder fenders and raked grilles, windshields, and A-pillars (Figure 96). The American influence on styling was unmistakable. Even entirely German firms such as Auto Union gave up the brand-specific identity on some of their products. They copied Detroit styling elements to compete more effectively in export markets, already a well-established practice at Ford of Cologne and Opel.

German auto makers felt compelled to increase exports because it was the only way to secure additional raw-material allocations. These were needed to increase production; raw materials for civilian industry were becoming increasingly scarce as a result of German rearmament. The draft was reintroduced in 1935, a four-year plan and establishment of dispersed Reich industries in 1936. At Opel, the craving for foreign exchange was so acute that the Kadett K 38 (1938–1940) was sold at dumping prices. In England, it cost £135, less than the equivalent price in Germany where it sold for 2100 Reichsmarks.

The average worker in the metalworking industries, who in 1938 grossed 220.44 Reichsmarks per month, had to set aside 9-1/2 months' pay to buy such a car. A married agricultural worker receiving payment in kind and a monthly gross of 124.25 Reichsmarks had to put away nearly 17 monthly paychecks, not considering cost of living. Even in the 1930s, car ownership remained a distant dream for the average wage or salary earner.

As a result of the forced earning equality imposed by the new regime, the average comrade in this new German Reich had no choice but to acquire a motorcycle, the second most important means of individual transportation in Germany and Europe behind the bicycle. Despite loud proclamations by the National Socialists, the lot of the private car owner had not improved. State-regulated wages were too low, and any increase in gross national product went directly to fulfill the State's insatiable appetite. Private consumption as a portion of net national product sank from 86 percent in 1932 to only 58 percent in 1938. Regardless, the masses had a sense of well-being, attributable to having come through the world economic crisis which for many represented only hunger, misery, and joblessness.

The National Socialist administration was aware of this situation, although it could not be discussed openly. In a speech at the IAMA, Berlin's International Automobile and Motorcycle Exposition of 1938, Hitler hid the realities behind a smoke screen and instead found a convincing argument to legitimize the creation of a *Volksauto*—a people's car. Among other things, Hitler declared "...that someday, we will arrive at the situation where our current auto production capacity is more than adequate to fill the possible demand, and a production increase is only conceivable if the market is opened to a much larger customer group, at a much lower income level. We are approaching this period of time, and therefore we will now begin construction of the mighty Volkswagen factory." (*Deutsche Kraftfahrt*, Hannover, Vol. 3, 1938)

As early as 1934, Hitler had challenged the Reichsverband der Automobilindustrie (Reich Automobile Industry Association, or RDA) to draw up plans for a Volksauto because, in his opinion, German industry was not offering such a people's car. However, the manufacturers represented in the RDA could not agree on the engineering layout of a future Volkswagen, nor were they interested in a product that competed with their own. As a way out of the dilemma, they decided to issue the design contract to an independent engineering consulting firm. Although engineers Josef Ganz and Edmund Rumpler were considered initially, their Jewish ancestry prohibited any further work with them. Instead, the RDA signed a contract with the Porsche design bureau.

Porsche chose the rear-engine layout favored by Josef Ganz, finally settling on the "optimum drivetrain layout" suggested by Béla Barényi in 1925 (Figure 110): an air-cooled four-cylinder horizontally opposed engine mounted longitudinally behind the rear axle, transmission ahead of the rear axle, differential between engine and transmission, and swing axles.

In 1936, as soon as Porsche presented his design drawings and calculations, the RDA gave the project back to the government. Angered by this move, Hitler handed the project to the DAF—the Deutsche Arbeitsfront, the German Workers' Front. Now, contrary to his original intentions, Hitler had put the NSDAP, his National Socialist German Workers' Party, into the car-making business. Following Hitler's will, the new car was named the KdF-Wagen, for the Nazi workers' society, *Kraft durch Freude*—Strength Through Joy.

Production startup for the KdF was slated for fall of 1939, with deliveries commencing in 1940. In that first year, planned production was 100,000 units and later 450,000 units. Considering the plant as it then existed and the number of employees, these numbers would not have been possible in any case.

The KdF car was to be made available only through an insured "savings plan." That is, cash purchases were not possible. Each week, prospective buyers had to paste at least a 5 Mark stamp into a savings book. Three full savings cards would buy an "inside steering" car—a sedan. However, the invasion of Poland in September 1939 triggered World War II and put off any civilian production of the KdF. By the middle of 1939, the Reich coffers had collected 110 million Reichsmarks. By the end of 1944, the government had amassed 268 million Reichsmarks from savers who in this way helped to underwrite the design costs, tooling, and building of the plant. Unfortunately, they never received their cars.

The Volkswagen project was by no means the only case in which the Party or its various organs interfered with the automotive economy. Examples include the DDAC, the German auto club, founded in 1933 with the "help" of the SA—Hitler's storm troopers, the *Sturmabteilung*—by consolidating all existing clubs. The 1939 Schell Program dictated which firms should build which cars. The Party, through various branches, also organized motorsports events and subsidized race car design and construction.

By far the most prominent "messengers of German know-how to distant lands" were race victories by the Mercedes and Auto Union Grand Prix cars, for "...never before has a nation achieved such a series of overwhelming victories and world records, as has Germany since

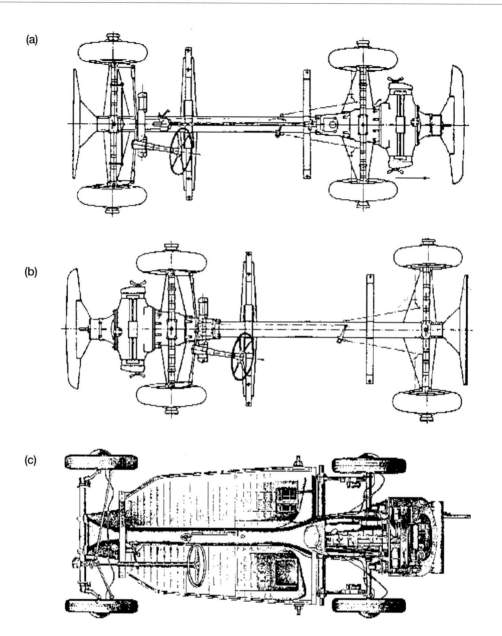

Figure 110. Barényi's "optimum drivetrain layout," 1925. Rear-mounted (a) or front-mounted (b) boxer engine, central tube frame with independent suspension by transverse leaf springs. For safety considerations, the rack-and-pinion steering box is mounted higher and farther back than usual practice, and frame-mounted bumpers are located front and rear. No car was ever built to these drawings. A Volkswagen chassis (c) of 1952, with platform frame (in effect, a central tube with welded-on sheet metal "wings"), boxer engine at the rear, independent suspension via torsion bars, a long and unarticulated steering column, and a steering box mounted above and ahead of the front suspension tubes. There are no frame-mounted bumpers.

1933" (*Motor Schau*, Vol. 2, 1937, p. 29). Indeed, German race cars were superior to their French and Italian opponents even on paper—streamlined, enclosed bodywork, independent suspension, larger and more powerful engines, extensive use of lightweight alloys, and, in the case of the Auto Union racers, an unconventional engine location ahead of the rear axle *à la* Rumpler. Until the outbreak of war, the racing teams of Bugatti, Delahaye, Alfa Romeo, and Maserati seldom had a chance to win, although a shortage of good drivers presented difficulties for Auto Union as well as Mercedes.

Originally conceived for the Wanderer Werke, the Porsche design bureau created a race car design meeting the 750-kg (1653-lb) racing formula of 1934. To better utilize the expected high power output, Porsche took the obvious technological leap in race car design (Figure 111). The engine location employed in the Auto Union Grand Prix car has been the state of the art in race car design since the 1960s.

Figure 111. Auto Union C Type, circa 1937. Designed by Porsche and continuously improved at Horch, under Eberan von Eberhorst, the V16 race car exhibits the layout advocated by Rumpler in 1921 (Figures 89 and 91), with engine ahead of the rear axle. This eliminated the driveshaft and lowered the center of gravity. Steel tube frame with Duralumin and aluminum bodywork, rear swing axle, and front twin trailing arms (crank arms, similar to the later Volkswagen), all suspended by torsion bars. V16 engine with Roots supercharger, 6010 cc, 520 hp (382 kW).

In contrast to the Auto Union GP car, the Mercedes racers (Figure 112) exhibited conventional layout with an upright front-mounted engine, ahead of the driver. As a result of sophisticated design (and despite limited technological superiority), Mercedes was able to score numerous wins.

Figure 112. Mercedes-Benz W 154, 1938. Conventional layout with front-mounted but inclined engine, allowing the driveshaft to run alongside the driver, lowering the driver's position and thereby lowering the center of gravity. Independent front suspension on A-arms with coil springs, rear suspension via de Dion axle with longitudinal torsion bars. High fuel consumption of 94 liters/100 km (2.5 mpg) necessitated large rear and central fuel tanks. For comparison, the Auto Union used 65 liters/100 km (3.6 mpg). V12 engine with two Roots superchargers, 2960 cc, 483 hp (355 kW).

Outside Germany, the fact that Mercedes and Auto Union received government subsidies was viewed with disdain; German superiority on the race track often was attributed to state sponsorship. However, the facts and figures speak otherwise. Although it is true that Auto Union did not take up Grand Prix racing until it had received assurances of financial support, this accounted for less than 1 percent of the annual turnover of the firm. Fully 75 percent of the cost of racing was paid by the Saxon firm itself; approximately 5 percent came from suppliers in the form of winner's bonuses. The remaining 20 percent came from the Reich in the form of victory prizes and bonuses ranging up to 10,000 Reichsmarks per Grand Prix victory, and a lump-sum grant of 300,000 to 450,000 Marks per year. The situation at Daimler-Benz was similar. Also overlooked is the fact that as of the mid-1930s, French race car constructors also enjoyed government support but without any significant success on the track.

Racing victories primarily served to promote the image of the German auto industry in foreign markets but had little effect on domestic sales. While Daimler-Benz lost market share between 1933 and 1939, the rather unsporty U.S. subsidiaries Ford and Opel made inroads on the market. Auto Union also improved its position, but only thanks to its DKW compact cars.

In addition to Grand Prix motorsports, the Reich also promoted motorsports for the masses. Usually under the umbrella of the National Socialist Motorist Corps (NS Kraftfahrer-Korps, NSKK), founded in 1931, it organized events in all regions of Germany, with regional to national significance. These include the three-day trials in the Harz Mountains, the Ostland-Treuefahrt (Ostland Loyalty Run), Brandenburgische Geländefahrt (Brandenburg Off-Road Trials), and the Mittelgebirgsfahrt (Highland Trials). Ostensibly these served to test series production or limited-production vehicles under difficult conditions, but their main purpose was to familiarize future soldiers with motor vehicle technology and off-road operations.

Commercial Vehicles and Buses, 1919 to 1939

After World War I, American manufacturers took the lead in the commercial vehicle sector. German trucks were no longer competitive, their technology antiquated. Promising prewar developments had been interrupted. After 1918, these could be implemented only after considerable delays or with the help of licensing.

For example, by 1913, shaft drive had replaced belt, pinion, and chain drive, at least on light trucks. Manufacturing costs, wear, weight, space requirements, serviceability, and noise emissions were all lower than those of other power transmission methods, and mechanical efficiency was higher. Yet chains and pinions held sway in German truck design well into the 1920s because Army management had specified these two designs for the "subsidy trucks" as of 1908. From the viewpoint of the military, this was perfectly understandable. During the war, raw materials came to be in ever shorter supply in Germany, and solid rubber tires had to be replaced by iron-tired wooden wheels or self-springing wheels. Chain drive, with the addition of some springs, turned out to be more elastic and wear resistant than shaft drive, which was incapable of withstanding severe shock loads. For this reason, some German firms even switched from shaft drive back to chains (Figure 113), but later had to return to shaft drive (*cf.* Figure 72).

German lack of progress in the commercial vehicle sector was equally evident in the pneumatic tire situation. In 1906, an American authority analyzed the German industry as follows: "...[they] have perhaps put forth the best efforts to fit pneumatic tires to motor 'buses. During the last year they have put out several types of motor 'bus tires...of a strength able to sustain the heavier 'buses even on country roads and at good speeds" (Pearson, Henry C., *Rubber Tires and All About Them*, The India Rubber Publishing Company, New York, 1906). Pearson could not yet report on Fritz Hoffman's synthetic rubber of 1908 nor industrial production of tires using synthetic rubber, which began in 1914.

Figure 113. NAG Type 07 brewery wagon, circa 1913. A three-tonner with rear-wheel drive via open chain. When NAG introduced driveshafts in 1919, the trade publication AAZ (Vol. 22/1919, p. 14) commented that "...during the war, inspiration by examples of foreign practice was lacking; only through magazines and captured vehicles did domestic industry learn of advances, particularly those made by American engineers." Four-cylinder engine, 8490 cc, 32 hp (24 kW).

After the end of the war, the situation was different. The Goodyear Tire & Rubber Company of Akron, Ohio, had taken the lead. As of 1918, the company produced serviceable tires suitable for heavy trucks and buses.

However, tires had limited load capacity and for the chosen 5-ton weight class, the necessary tire diameter was so great that they created a formidable height difference between truck beds and loading ramps. Even Goodyear's proposal to distribute the load across a pair of tandem axles with correspondingly smaller tires did not find favor among freight operators. To amortize development costs, at least over the long term, the tire and components manufacturer entered the ranks of truck manufacturers, introducing its Goodyear truck in 1920 with pneumatic tires on a rear tandem axle (Figure 114). Goodyear's coast-to-coast publicity runs convinced American trucking firms and commercial vehicle manufacturers. Increasingly, they chose pneumatic tires made by Goodyear and other American tire manufacturers.

One of these, B.F. Goodrich, also of Akron, signed an agreement with the Continental Caoutchouc and Gutta Percha Company of Hannover, enabling the latter to be the first firm to produce pneumatic tires for commercial vehicles in Germany (1921). Other German firms soon followed, and by 1923 six German manufacturers had trucks up to 4 tons in their programs. In the same year, Büssing brought out a three-axle omnibus, Europe's first large bus with pneumatic tires (Figure 115). This was followed in 1927 by a three-axle double-decker bus for Berlin's ABOAG—the Allgemeine Berliner Omnibus-Aktien-Gesellschaft. By 1928,

Figure 114. Goodyear 5-ton, 1921. With the sole purpose of marketing its pneumatic tires, Goodyear entered the ranks of commercial vehicle builders with this 5-tonner. One unusual feature was its tandem rear axles, in 1920 still driven by chains, soon to be replaced by a driveshaft. Otherwise, designer Ellis W. Templin used components bought from outside suppliers. In 1926, when production of pneumatic tires for commercial vehicles finally exceeded that of solid-rubber tires, Goodyear terminated its truck production.

eleven additional German commercial vehicle manufacturers began producing three-axle vehicles with pneumatic tires, which became a German specialty led by Büssing. In 1930, the firm claimed a domestic market share of 90 percent and more than 50 percent worldwide.

Except for the diesel engine and electronics, no other technical system has had as lasting an impact on the importance of commercial vehicles as pneumatic tires. This advance brought with it considerable technical and economic benefits. Because pneumatic tires could absorb small road irregularities, formerly over-dimensioned frames, springs, and drivetrain components could be made lighter. The resulting lower weight increased payloads and speeds.

Figure 115. Büssing III GL 6, 1923. Limited by tire load-carrying ability, higher payloads mandated four rear wheels. Following Goodyear's lead, Büssing also adopted a tandem layout, with one tire on each end of an axle. This is the first European full-size bus, designed by Paul Filehr, with individual drive and rotating balance-beam suspension for both rear axles. Four-cylinder engine, 7850 cc, 50 hp (37 kW).

Quite possibly, the economic effects were even greater. Pneumatic tires were gentler on roads, bridges, cargo, and passengers. Iron- and solid-rubber-tired vehicles had been restricted to factory grounds and the immediate local area; now, trucks and buses could range over greater distances. Suddenly, the railroads had serious competition. The fare and tariff wars between road and rail had begun, and its effects are still felt today.

Hydraulic brakes suitable for ordinary service also originated in the United States. Because an 1895 patent by Hugo Mayer of Rudolstadt for a "hydraulically actuated brake for [horse-drawn] wagons" had been forgotten or overlooked, the German automotive industry had to make use of the brake system developed by Malcolm Loughead (later Lockheed). Alfred Teves acquired the German rights to the Lockheed system in 1926. The first production passenger car with Lockheed brakes was the Chrysler 70 of 1924; the Triumph 13/30 of 1925 was the first British car so equipped. Germany's Adler Standard 6 followed in 1926 (Figure 80). In the German commercial vehicle sector, Krupp was the first in 1926, Büssing in 1927, and the Mercedes Lo 2000 (Figure 123) in 1931, along with MAN, Henschel, and Saurer.

Among trucks and buses heavier than 3 tons, the limits of purely hydraulic brakes were soon reached. To this day, their use remains limited to light trucks. Following the example of railway brakes, Knorr Bremse AG introduced its air brake in 1923, still standard equipment on heavy trucks. Railway use of air brakes dates back to George Westinghouse's 1869 patent.

Other early inventions did not prevail. Conventional transmissions had several major drawbacks: jerk and gear clash during shifts, power interruptions, and, not uncommonly, gear failure. Engineer Fritz Mayer of Lippische Werke Detmold (LWD) in 1923 attempted to circumvent these with his clutchless transmission that employed oil pressure for gear shifts. Air suspension, designed by Fritz Faudi and offered by Rheinmetall in 1926, did not reappear until 1958 on buses, and even later on trucks and semitrailers.

Trucks and buses with front-wheel drive remained exceptions, at least in Germany. In 1923, LWD/Fritz Mayer and its Swiss licensee, SLM, built a few municipal vehicles. Front-drive buses by ABOAG built to patents by Richard Bussien/Voran (1928) and trucks by Edmund Rumpler (1931) did not go past the prototype stage. In the United States, the Aeromarine Plane & Motor Co. of Keyport, New Jersey sold approximately 30 of its Uppercu single- and double-decker front-drive buses. The only maker to achieve significant numbers was George Latil, France's pioneer of front-drive commercial vehicles. From around 1900, he built front-drive trucks and tractors, switching to all-wheel drive in the 1930s.

Nevertheless, front-drive commercial vehicles were able to carve only a few specialized niches in the high weight classes, even after World War II as homokinetic or constant-velocity joints only gradually replaced universal joints, which were only marginally suitable. Front drive without the benefit of power-assisted steering, the only configuration known at the time, did not seem particularly necessary, apart from a few special-purpose vehicles and city and airport buses.

Delivery vehicles were another matter. After their front-drive three-wheeler, Tempo and Framo introduced a four-wheeled front-driver in the 1930s, followed by Citroën in 1939 (Figure 116) and Chenard et Walcker in 1941. Both French firms employed forward-control cab designs, hydraulic brakes, independent front suspension, and lowered cargo beds—all basic design principles of modern delivery vans.

Figure 116. Citroën TU, 1939. A forward-control fourgon with unit body construction, sliding cargo door on the right side, and, for some models, a loading ramp for small animals on the left side. Front drive permitted a low, flat floor. Intended as a delivery van for commercial enterprises, during the war the Citroën mainly served as a military ambulance. A total of 1,749 examples were built before production ended in 1942. Four-cylinder engine, 1628 cc, 35 hp (26 kW).

Formerly specializing in trucks, in 1921 Fageol Motors Co. of Oakland introduced a bus that is now regarded as the industry's first step away from its heritage in truck engineering and toward independent bus development. With its Safety Bus, then its 22-seat Safety Coach (Figure 117), the Fageol brothers abandoned the high-framed truck design ethic; the low-slung frame, wide track, and lowered coachwork emulated key design features of the American automobile. With their success in overland and intercity service, Fageol buses found imitators in the United States and in Germany. Bus manufacturers now had to take the intended service into consideration: low-frame chassis for city and multi-purpose buses, taller frames from truck production for overland and sightseeing buses (Figure 115). The average height of German single-decker buses fell from 2.80 meters (9.2 ft) to 2.50 meters (8.2 ft) or less—in other words, providing improved handling characteristics and less danger of tipping.

Figure 117. Fageol Safety Coach "Parlor Type," circa 1927. Introduced in 1921, the Fageol Safety Bus, with its low-slung frame, low silhouette, and general appearance of a large sedan, influenced bus design for decades. The floor was only 55.8 cm (22 in.) above ground level and was reached by only two steps. In the de Luxe edition, each passenger compartment, with transverse bench seating, had its own door. Shown here is a "club car" version, with wicker seats attached via suction cups. Choice of four- or six-cylinder Scott & Hall OHC engines.

At virtually the same time, firms new to the trade introduced all-steel construction to the bus industry. In 1924, the Six Wheel Co. of Philadelphia produced the first bus with coachwork, with the exception of the roof, made entirely of steel. The Six Wheel Co. was the sales arm of the American Motor Body Corp., a subsidiary of Bethlehem Steel. In Europe, the Waggonfabrik Uerdingen served a similar pioneering role. In 1925, Uerdingen built an all-steel body on a Büssing three-axle chassis. By 1931, approximately 70 percent of all German buses had metal bodies—most of steel, some of aluminum alloy. In this sector at least, the bus industry was 15 to 20 years ahead of the auto industry. Advantages over the wood-framed construction inherited from horse-drawn coach practice included lighter weight, greater safety and longevity, simpler repair, and faster production.

All-metal construction represented the necessary precursor to unit bodywork. This was first shown in production-ready form in early October 1927 at the American Electric Railway Association (AERA) conference in Cleveland. The three buses exhibited at the 1927 AERA gathering were forward-control unit-body structures by Twin Coach, ACF, and Versare, with styling and technical cues borrowed from American streetcars (trolleys).

Although Büssing imported a Twin Coach to Germany in 1930 and in 1931 developed the first so-called Trambus in cooperation with the Hannoverschen Waggonfabrik, neither these nor any of the other established bus manufacturers adopted the American proposals. Only outsiders who did not need to consider their existing concurrent truck production built buses with unit bodywork: Waggonfabrik Uerdingen in 1930–1932, Maschinen- und Fahrzeugfabrik F. Stille of Münster in 1930, Triebwagen- und Waggonfabrik (Railcar and Wagon Factory) Wismar in 1933, and the Harmenig coachworks in 1939 and 1942 (Figure 118). General German bus practice did not embrace unit bodywork and longitudinal rear-mounted engines until the 1950s. Transverse rear engines, such as those built by Twin Coach and Yellow as early as 1934–1935, came even later.

Figure 118. Ford BB, 1939–1942. Josef Deiters' masterpiece. Stiff plywood panels made of "Delignit"—from Deiters and lignum *(Latin for wood)—took the place of sheet metal roof and side panels. The frame consisted of enclosed box sections, their tops sealed by the Delignit floor. Thanks to Deiters' lightweight rigid design, buses built by the Harmenig coachworks weighed no more than gasoline-powered models, despite their 700 kg (1543 lb) weight penalty of an Imbert gas generator at the rear and firewood stored on the roof. Four-cylinder engine, 3285 cc, 52 hp (38 kW) when operating on gasoline.*

Construction of the Reichsautobahnen influenced bus development. When the first segments were opened in 1935, the German Reich Railways instituted bus passenger express service between Frankfurt and Darmstadt (Figure 119). These were thoroughly conventional chassis with separate frames but with streamlined forward-control bodies and the engine located in the cabin next to the driver. In the planning stages, or already existing as prototypes, were Autobahn giants with three- or four-axle chassis and underfloor engines of up to 350 hp. The Schell Program and the outbreak of war halted any further development.

Forward control and underfloor engine layout established themselves around 1935 as hall-marks of modern commercial vehicle design. Even earlier, these principles had been applied by Orion of Zurich, Switzerland, 1903–1910; by Hanomag/Krupp/Gaggenau, as licensees of

Figure 119. Mercedes-Benz Lo 3100, 1935. After the first sections of the Autobahn were completed in 1935, the Deutsche Reichsbahn (German Reich Railways) inaugurated high-speed bus service using somewhat bulbous forward-control vehicles on conventional chassis. The open folding top leads us to conclude that the freedom-loving workers of the Reich were being offered traveling enjoyment rather than speed. Both bus and rail, such as the Rheingold Express between Holland and Switzerland, averaged nearly identical speeds of approximately 70 km/h (43.5 mph). Six-cylinder diesel engine, 5530 cc, 75 hp (55 kW).

Stoltz, 1905–1908; and by others. It was not European firms, but American manufacturers Twin Coach (1927), Delling (1928), and White (1932–1935) that resurrected the theme, which must have captured the attention of Paul Arendt, chief engine designer at Büssing. This would allow eliminating the nose of the vehicle; allowing a larger load bed while maintaining the same footprint as a front-engined vehicle; better forward visibility; easier maneuvering; lower center of gravity; less noise, odor, and heat radiation in the driver's cab; shorter driveshaft and fewer or no driveshaft bearings; utilization of the "dead" space under the bed; and often easier access and simplified removal of the engine and transmission.

Because he did not find any sympathetic ears at Büssing, Arendt developed a "flatnosed" underfloor-engined truck for Hanomag (Figure 120), which was built in limited numbers from 1933 to 1936. Meanwhile (1934), Arendt had returned to Büssing, where he developed an underfloor diesel engine. As of 1936, this was installed in two- and three-axle tram buses. Büssing therefore is regarded as the firm that popularized underfloor-engined buses and trucks.

The engines described here were all diesels. In the early 1930s, at least in Germany, there was hardly a new engine design that was not based on Rudolf Diesel's principle, as we will see in the next section.

Figure 120. Hanomag HL, 1933. Although the principle was known since 1896 (Figure 62), underfloor engines and forward control did not gain acceptance until the 1930s. The first full-size German commercial vehicle with these features was the Hanomag HL, in which the normally upright engine was canted 90 degrees to the left and mounted below a conventional ladder frame, slightly behind the left front wheel. The HL also was available as a bus. Four-cylinder diesel engine, 5193 cc, 55 hp (40 kW).

The Diesel Engine Arrives on the Automotive Scene

Rudolf Diesel (1858–1913) was one of the great engineers responsible for molding the internal combustion engine as we know it today. He had attempted to put into practice the theories proposed in 1824 by French theorist Léonard Sadi Carnot (1796–1832), in which an ideal engine must operate with the greatest possible differences of temperature and pressure to improve thermodynamic efficiency (see Appendix, Carnot cycle). However, after Diesel applied for a patent for his "new rational heat engine" in 1892, experiments at the Maschinenfabrik Augsburg confirmed Diesel's nagging afterthought: Carnot's theory could not be applied to practice using the methods and knowledge of that time. Regardless, he had developed the process that today bears his name. It is characterized by intake of air, without fuel, followed by compression (at that time to approximately 30 bar—440 psi). The heat of compression would ignite the introduced fuel without need of a separate ignition system (self-ignition).

In 1897, after many years of experimenting and several setbacks, Diesel and the Maschinenfabrik Augsburg, later MAN, finished their third experimental engine, the first to produce useful power. This "first diesel engine" is preserved in Munich's Deutsches Museum. The stationary single-cylinder engine had a displacement of 19,635 cc, produced 18 hp at 154 rpm, and, with a thermal efficiency of 26 percent, considerably exceeded that of a steam engine by approximately 10 percent and gasoline engines for motorcars by approximately 15 percent.

After Diesel's basic patents expired in 1908, many firms began to manufacture diesel engines. Marine diesels were built as of 1903, and diesels for service in submarines as of 1904. It appeared that lightweight diesels for road vehicles were in the foreseeable future.

This did not come to pass because of the difficulty in introducing fuel into the cylinder. Fuel, held by an injection valve, had to be injected into the cylinder at the right instant by a compressor. However, this compressor system increased weight, size, and manufacturing costs, and it reduced mechanical efficiency. It was not suitable for high engine speeds nor for rapid load changes, as would be expected in a road vehicle.

Diesel himself and some firms also had, on many occasions, unsuccessfully attempted to replace air injection, suitable for stationary engines, with high-pressure injection. It was Vickers in England who, in 1910, began the development of the solid-injection pump system. As of 1913, Vickers started production of compressor-less marine and stationary diesel engines.

Inexplicably, German diesel engine manufacturers stuck to air injection—thereby stultifying development of lightweight automotive diesels. Necessity being the mother of invention, these would be developed in the 1920s in response to the economic and political situation of postwar Germany. Reparations and the loss of vital coal fields put benzol in short supply. Mineral oil could no longer be obtained from Eastern Europe but had to be bought at inflated prices from the United States.

In 1919, the German magazine *Der Motorwagen* (Vol. 29, p. 542) bemoaned the immediate postwar situation: "The present shortage and usurious prices of benzol force motorized plow designers to work toward use of oil-burning engines." The firm of Benz-Sendling Motorpflüge ("motor plows") GmbH was founded in June of the same year and rose to the challenge in 1922 with a diesel tractor that did not employ the air injection principle (Figure 121). This was followed by MWM in 1923 with its road tractors "Motorpferd" ("Motor Horse") or "Eisenpferd" ("Iron Horse"), as well as Benz and MAN in 1924 (Figure 122) with the first diesel trucks. As a positive outcome to its dire economic situation, Germany had taken the lead in automotive diesel engine design.

As powerplants for these Benz and MAN trucks, diesels had finally broken into the automotive field; this could easily have been accomplished a decade sooner, thanks to McKechnie. When Bosch introduced its diesel injection pumps, governors, and nozzles in 1927, as market-ready products based on the Acro pumps designed by Franz Lang, other firms in Germany, Britain, and Switzerland took up construction of diesel trucks and buses.

Bus operators counted on lower operating costs with diesels. The Associated Equipment Co. (AEC) developed an "oiler" (the conveniently patriotic British euphemism for diesels) as early as 1928. These were delivered to the London General Omnibus Co. (LGOC) beginning in 1931. The Berliner Verkehrs-Aktien-Gesellschaft (BVG, Berlin Transit Corporation) commenced operations with the first diesel-powered double-deckers made by Büssing and Daimler-Benz in 1934.

Figure 121. Benz-Sendling Type S motor plow, 1922. In 1909, the Motorenfabrik München-Sendling, under its founder Otto Vollnhals, built the first German agricultural tractor of modern configuration. Benz, however, as a Sendling licensee, adopted the "motor plow" concept instead. The Benz-Sendling is characterized by a single large drive wheel and, most important, a compressor-less diesel engine, making it a pioneering design in the history of diesel vehicles. Further development produced a four-cylinder Benz diesel for commercial vehicles as of 1924. Two-cylinder diesel, 5723 cc, 30 hp (22 kW) at 800 rpm.

Figure 122. MAN diesel truck, 1924. In early 1924, Benz presented a truck powered by a compressor-less diesel engine. MAN followed a few months later. While Benz chose a prechamber concept, based on designs by Prosper l'Orange, MAN took the direct-injection route. MAN had developed the stationary diesel engine concept into functional powerplants and in 1915 began building trucks with gasoline engines licensed from Saurer. Four-cylinder diesel engine, 5650 cc, 40 hp (29 kW) at 900 rpm.

135

These buses were massive hood units with six-cylinder engines that turned a maximum of 1600 rpm. Compared to gasoline engines, diesels had greater difficulty using high engine speeds to compensate for the power loss expected with smaller displacement. For this reason, five-ton trucks were regarded as the lower limit for diesel vehicles into the 1930s. However, when Daimler-Benz introduced a small diesel two-tonner in 1932 (Figure 123), capable of turning 2000 rpm, the automotive press wrote glowing editorials in expectation of an imminent diesel passenger car.

To an even greater extent than the general public, automotive and diesel authorities had long speculated on the possibility of diesel cars. As early as 1896, Maschinenbau-AG Nürnberg, the other predecessor to MAN, had worked on diesel engines for carriages and railcars. Ludwig Lohner of Austria attempted to run a diesel car for the 1898 racing season. In 1908–1910, under contract to Rudolf Diesel, Heinrich Dechamps designed a truck engine for the Swiss automobile firm SAFIR of Rheineck. None of these efforts met with success. One

Figure 123. Mercedes-Benz Lo 2000, 1932. Because they cost 3500 to 4425 Marks more than gasoline engines, diesels only gradually gained acceptance. Only after Mercedes introduced its lightweight Lo 2000, with a diesel option of only 1260 Marks, did the proportion of diesel trucks and buses increase. The two-tonner was also available with a gasoline engine and various bodies, including buses. Four-cylinder diesel engine, 3770 cc, 55 hp (40 kW) at 2000 rpm.

minor exception was Hermann Dorner, whose Hannover-based Dorner Oelmotoren-AG (Dorner Oil Engines Inc.) built approximately 10 cycle cars in 1923–1924 (Figure 124) with single- or two-cylinder diesel engines. Eugen Diesel described the Dorner car as follows: "...of all the world's motor vehicles with any chance of becoming a true people's car, the Dorner had the lowest operating costs. It was most likely also the world's first diesel passenger car." (*Die Geschichte des Diesel-Personenwagens*, Stuttgart, 1955, p. 61)

In the 1930s, after the arrival on the market of the Bosch injection system, diesel passenger car development advanced along a broad front. From 1932, the British engine manufacturer Gardner converted approximately 40 gasoline-powered cars to its own prechamber diesel but was unable to convince any auto manufacturer to produce diesel cars in volume. In 1933, Perkins offered 40- and 70-hp diesel engines for passenger cars.

Figure 124. Dorner heavy oil car, 1924. Aircraft engineer Dorner (not to be confused with Dornier) had a unique answer to the question of the day, which was how "one could exist alongside Ford and engage in extensive development work." His solution was an extreme curtailment in the amount of material used to produce a car, and cheaper fuel (i.e., diesel microcars). The result was a cycle car with an admittedly simple frame but featuring rear-wheel drive via a driveshaft and differential, and four-wheel brakes. Its most interesting component, however, was an air-cooled diesel engine with an injection system about the size of the palm of a human hand, located above each injection nozzle. V2 diesel engine, 663 cc, 5 hp (3.7 kW) at 1300 rpm.

In 1934, the Maschinenfabrik Oberhänsli of Bregenz (Austria) expanded its commercial vehicle engine line with a 2.2-liter powerplant, followed a year later by a 1.5-liter for installation in passenger cars. In 1935, Saurer, which assembled Chrysler products in Switzerland, fitted two Dodges and a Plymouth with six-cylinder direct-injection diesels. One Dodge found its way to the 1936 New York Auto Show; the other two operated for years in Switzerland. In 1938, Studebaker Distributors Ltd. of London offered Canadian-built Studebakers with British-built Perkins diesels.

Public acceptance was lukewarm, although the automotive press (at least in Germany) repeatedly pointed out the risks to the fuel supply posed by foreign exchange rates and economic and political concerns. It had been shown that neither the engine manufacturers nor their subsidiaries or assembly plants were in any position to implement capital-intensive investment or marketing measures of the caliber required for an alternative powerplant. A more recent example is the case of the Wankel engine in the 1960s.

Of necessity, these manufacturing and marketing tasks fell on the auto manufacturers themselves. On November 27, 1934, Citroën presented its diesel passenger car Type 10 série Di, which except for its engine (four-cylinder, 1766 cc, 33 hp) was a Type 10 NH with four doors, six side windows, long wheelbase, and a glass partition between driver and rear passengers. This was the result of a 1933 contract between SA André Citroën of Paris and Harry Ralph Ricardo of Sussex for the development of lightweight diesel engines. In the spring of 1935, Citroën gave 100 prototype examples of the Type 10 série Di (Figure 125) to selected drivers. These probably consisted of 25 taxis and 75 Commerciales. The Commerciales resembled the sedans, except for a horizontally split tailgate.

In November 1935, Daimler-Benz began series production of its diesel passenger car, the 260 D, which had not yet been shown to the public. Daimler finished 13 examples by the end of 1935. Daimler-Benz formally presented the Mercedes 260 D (four cylinders, 2545 cc, 45 hp) at the Berlin IAMA in February 1936, as a six-seat landaulet taxi (Figure 126) derived from the 200/230 series.

Also shown at the 1936 IAMA was Hanomag's 1.6-liter diesel engine, which the firm offered as a complete unit for installation in truck tractors, trucks, and delivery vehicles. As of midyear, it hoped to offer the engine in its midrange car, the Rekord. Indeed, a two-door Rekord sedan fitted with a diesel engine (and automatic clutch) graced the Paris Salon in the fall of 1936. However, the 1.6-liter engine then vanished (temporarily, as it turned out) because Hanomag had not been able to make it reliable in daily service. Enlarged to 1.9 liters, the engine made it into production the following year. In the small tow vehicle SS 20 and the agricultural tractor RL 20, both of which went into production in 1937, the engine developed 19.8 hp. As of 1938, it produced 35 hp in the Rekord passenger car (Figure 127). In February 1939, Hanomag established four international records for diesel vehicles, using a streamliner designed by Koenig-Fachsenfeld on the new Autobahn near Dessau. These records remained unbroken until the 1960s.

Figure 125. Citroën Type 10 série Di, 1935. The world's first production diesel passenger car was developed with the aid of Ricardo and his British patent number 384,340. Citroën presented the car in November 1934 and began deliveries in the spring of 1935 with approximately 100 units. Shown here are three diesel-powered 10 CV sedans at Montlhéry, July 1935, with Alexander Ferguson (wearing hat), Ricardo's chief engineer, and René Wilsner (second from right), head of the Citroën diesel department. The Citroën diesel engine was also installed in delivery vehicles. Four-cylinder diesel engine, 1766 cc, 33 hp (24 kW) at 3000 rpm.

Figure 126. Mercedes-Benz 260 D, 1935. Deliveries began in November 1935, with 13 examples shipped by the end of the year. The official presentation took place in February 1936 at the Berlin IAMA. Choices included a Pullman sedan and a landaulet, initially with a frameless (as shown here) and later with a framed rearmost side window. The Pullman was priced at 6800 Reichsmarks; the landaulet was priced 50 Reichsmarks less. Four-cylinder diesel engine, 2545 cc, 45 hp (33 kW) at 3000 rpm.

Figure 127. Hanomag Rekord Diesel D 19 A, 1938. After experiments with 1.5- and 1.6-liter diesel engines beginning in 1933, during 1937 designer Laska Schargorodsky was able to make a larger, 1.9-liter prechamber diesel ready for production. It was presented in a Rekord sedan at the February 1937 IAMA, and a reduced-output version went into production that same year powering a small tow vehicle and an agricultural tractor. It was fitted to the Rekord passenger car in 1938. The two-door sedan was priced at 4475 Reichsmarks. Four-cylinder diesel engine, 1909 cc, 35 hp (26 kW) at 3000 rpm.

With these two models, the end of the 1930s saw two mass-produced diesel passenger cars on the European market: the Mercedes 260 D (1935–1940, 1,967 units produced), with emphasis on taxi models; and the Hanomag Rekord (1938–1940, 1,074 units produced), available only as a personal car, not a taxi. The Hanomag, in four-door sedan form, was priced at 4875 Reichsmarks; the Mercedes cost 6800 Reichsmarks. The lowest-cost model was the Pullman Limousine Kraftdroschke (Power Taxi), which Mercedes offered at the highly competitive price of 6750 Reichsmarks.

And what of Citroën? At the pioneering French firm, passenger car diesel development halted abruptly after Michelin acquired a controlling share of Citroën stock in 1935 and apparently had no interest in diesel cars. At Daimler-Benz, on the other hand, the 260 D was eventually made available to private customers, with the result that a continuous evolution of diesel car technology was established in Stuttgart, an evolution that continues today. Hanomag did not resume passenger car production after World War II.

In the United States, stationary diesel engines were offered by Alexander Winton as of 1913 and by Clessie Lyle Cummins as of 1919. However, thanks to low gasoline prices in the United States, there was little interest in diesel vehicles. Serious diesel engine development did not get underway until the 1930s, when General Motors, Waukesha, Cummins, and Caterpillar brought out two- and four-stroke diesels for trucks, tractors, and construction equipment.

Clessie Cummins (1888–1968) is regarded as the father of the American diesel truck. Prior to the stock market crash of 1929, he built small diesels for marine applications and earth-moving equipment. Orders dried up after the crash; therefore, to save his firm from collapse, Clessie initiated a spectacular public relations campaign. In it, he shoehorned a four-cylinder marine diesel under the hood of an aging Packard and covered the 1275 km (792 miles) from Indianapolis to the New York Auto Show with only $1.38 in fuel costs. With the help of the Associated Press, "overnight, he made the entire country aware of the diesel" (Diesel, Eugen, *Die Geschichte des Diesel-Personenwagens [History of the Diesel Passenger Car]*, Stuttgart, 1955, p. 69). Because he had no suitable engines ready for delivery and to remain in the headlines, Cummins undertook diesel record runs at Daytona in 1930 and 1931 (Figure 128). Orders began to come in, and Cummins began production of diesel engines for trucks—but not for cars. The United States was not yet ready for diesel passenger cars.

Figure 128. Cummins-Duesenberg diesel race car, 1931. In early 1931, Clessie Cummins and driver Dave Evans did record runs on Daytona Beach, followed by the Indianapolis 500 in May. Their Duesenberg racer was powered by a tuned Cummins Model U diesel engine, which normally produced 50 hp at 900 rpm. Driving the same car that was fitted with a top, windshield, luggage boxes, and headlights, Clessie and his financial director William G. Irvin visited, among others, André Citroën in Paris, who had an interest in Cummins diesel engines. Four-cylinder diesel engine, 4700 cc, 86 hp (63 kW) at 1700 rpm.

This reserve on the part of the American market had an effect on production and market statistics in Germany: neither Adam Opel AG, part of General Motors since 1929, nor Ford in Cologne offered any diesel commercial vehicles before the end of World War II. Opel, the biggest maker of commercial vehicles and passenger cars in Germany in the 1930s, relied on its gasoline-powered 2.5/3-ton Blitz. Ford offered its 1.75/3-ton truck with a choice of four-cylinder or V8 engines. Together, the German-American firms accounted for 50 percent of truck production. When smaller manufacturers are included, Opel and Ford's share represents 80 percent of the gasoline truck market up to 3 tons. Only above 3.5 tons did diesels dominate, with a share of 85 to 97 percent (1938 statistics, rounded off, from *RDA Tatsachen und Zahlen,* "RDA Facts and Figures," Berlin, 1939, pp. 47–48).

Diesel passenger cars had to content themselves as bit players until around 1974. In terms of smoothness, odor, noise, weight, and power, they could no longer compete with gasoline engines. Their numbers were so low (see Mercedes and Hanomag passenger cars in Figures 126 and 127, respectively) that they did not even appear in the statistics.

Motorized Forces in World War II, 1939 to 1945

The title of this section invites the assumption that land-based conflict in World War II always involved engagements between opposing motorized vehicles and mechanized units. This, however, is not the case.

Technological advancements in small arms and light weapons such as the infantry rifle, machine gun, and automatic weapons, and more remotely in heavier weapons such as armor, aircraft, and rockets, finally eliminated the horse as an instrument of war. Nevertheless, horses and mules played a role that should not be underestimated—not in cavalry operations but rather as draft animals, primarily on the Eastern (Russian) Front.

German Army cavalry units were disbanded in November 1941, but smaller SS mounted units, SS cavalry regiments, and an SS cavalry brigade, established in 1940, continued their independent existence. As late as March 1943, individual Cossack units were consolidated into a Cossack cavalry regiment that fought against the Soviets on the German side, mainly to combat partisans.

In the first cavalry attack of World War II, on September 1, 1939, a regiment of Polish Uhlans drew their sabers and charged against German infantry, only to be decimated by a hail of fire from tanks and motorized units that had been masked by a curve in the road. In the last cavalry engagement involving German forces, on September 23, 1939, machine gun positions again determined the outcome of the skirmish, as Uhlans, pursuing retreating German cavalry, rode into their crossfire.

The German Army entered the war with 573,000 horses; a total of 2,750,000 solipeds served in the Wehrmacht, three-quarters of them in the East. When snow drifts, the bottomless mud of Russian roads, and engine blocks cracked by bitterly cold temperatures forced truck transport to a standstill, horse-drawn wagons and sleds often formed the only link between the front line and supply depots. Approximately 60 to 63 percent of the animals drafted for the war—

around 1,650,000 animals—died "for Führer and Fatherland." Their average life expectancy was four years; that of a truck normally about a year, and by early 1945 only seven weeks.

It probably will never be known how many horses were lost in combat on the Soviet side. With 3.5 million horses "under arms," the Soviet Union was the greatest cavalry might of World War II. At least in this respect, displacement of these mistreated and abused creatures by advances in weapons systems and mechanized transport indeed represents progress.

As a "local means of transport," the horse was also not suitable for conquering combat theaters of continental dimensions in the time available. More than ever, World War II was a war of high mobility. Hundreds of thousands of vehicles took part and determined the course of the war. German numerical superiority in military vehicles at the outset of the war eventually led to a bewildering mixture of vehicle models and types. At the front, the sheer variety of supplementary vehicles requisitioned from the private sector, captured vehicles, and material shortages hampered both routine maintenance and quick repairs. Having the right spare part at the right place at the right time was a matter of pure luck.

Long before the outbreak of war, the government had instituted measures to avoid precisely this deficiency. Its measures consisted of both economic and technical intervention.

Beginning in 1935, German rearmament and its associated raw material shortages spawned a "four year plan" announced in October 1936. This emphasized technical innovations, particularly those related to extraction of and applications for domestic natural resources and materials. This four-year plan was short-lived. In the summer of 1938, the plan was replaced by the "New Production Plan for the Defense Economy," its provisions tailored to satisfy the interests of IG Farben. The autocratic policies of the government advanced the cause of synthesizing raw materials, although these raw materials, diesel fuel, and gasoline were actually available at lower cost on the world market.

In 1936, the Reich Transportation Ministry announced tax benefits for commercially available trucks with limited off-road capability. This had an immediate effect on technology and product lines, resulting in proliferation of all-wheel drive models. Büssing, for example, offered its 4.5-ton truck in two versions—500 S (for "steuerbegünstigt"—tax advantage) and 500 A (for Allrad, all-wheel drive).

In March 1939, the General Deputy for Motor Vehicles, Adolf von Schell, issued a decree restricting vehicle and component types, specifying which and how many types and models were to be built by which firms. Büssing, for example, was required to cease production of its underfloor models. Because of the unexpected early onset of the war, the Schell Program could not be implemented in its intended scope.

In contrast to previous subsidy policy, which was limited to issuing specifications and requisitioning civilian trucks when and if needed, the National Socialist government equipped its three branches of service—Wehrmacht, Luftwaffe, and Kriegsmarine—with their own vehicles that would have to fill expanded roles in the event of future military conflict. As a result, commercially available trucks were joined by vehicle designs specially conceived for unusual duty. These included, among others,

- All-terrain and amphibious vehicles based on the KdF/Volkswagen, 1939–1945 (Figure 129)

- Light, medium, and heavy Einheits-PKW (standard passenger cars), 1936–1943 (Figure 130)

- Einheits-Diesel (standard diesel), 1937–1940 (Figure 131)

- Krupp-Protze (artillery tractor), 1934–1942 (Figure 132)

- Armored scout cars, both wheeled and tracked, 1930–1945

- Halftracks, 1934–1945 (Figure 133)

Only months before the outbreak of war, Porsche received a contract to develop an all-terrain vehicle. He chose the KdF car as its basis. He raised the suspension and exchanged the enclosed bodywork for an open structure (Figure 129). The low weight of the Kübelwagen, its generous ground clearance, and rear engine location compensated for its lack of all-wheel drive, a feature found indispensable on the heavier Einheits-PKW (standard military passenger cars). Along with the amphibious Schwimmwagen, introduced in 1941, approximately 69,000 examples left the assembly line. With the American Jeep, these were among the most reliable and, in terms of maintenance, least demanding vehicles of World War II.

With an eye toward improved off-road capability and rationalization of maintenance, warehousing, and supply, the Heeres-Waffenamt (HWA, Army Ordnance Office) developed the so-called Einheits-Fahrgestelle (standard chassis) for passenger cars, in which engines of various manufacturers were installed.

The HWA laid out standards for a light, a medium, and a heavy car design. All had all-wheel drive and locking differentials; some had all-wheel steering. Because of their high weights, complex construction, and high maintenance requirements, neither the light nor the heavy versions were successful. Only the medium chassis, with a Horch V8 (Figure 130) or more rarely with an Opel six-cylinder engine, fulfilled expectations. Approximately 12,000 medium units were produced, compared to approximately 4,800 cars on heavy chassis and 14,200 on light chassis.

In contrast to the standard passenger cars, the standard truck—also known as the Einheits-Diesel (standard diesel) (Figure 131)—achieved a higher degree of simplification insofar as the

Figure 129. Volkswagen Kübelwagen, 1940–1945. Porsche's designs for the Type 62 and Type 82 Kübelwagen, officially designated Kfz.1 le.gl.PKW 4×2—military shorthand for "leichter geländegängiger PKW mit zwei angetriebenen Rädern" (lightweight all-terrain passenger car with two driven wheels)—were largely similar to the KdF car, intended for mass production and eventually familiar as the Volkswagen Beetle. The word "Kübelwagen" is derived from "Kübelsitz-Wagen," an open, often doorless passenger car with bucket ("kübel" means "bucket") seats. Rear-mounted four-cylinder horizontally opposed ("boxer") engine. Type 62: 985 cc, 23.5 hp (17 kW). Type 82: 1130 cc, 25 hp (18 kW).

Figure 130. Horch EFm Einheits-PKW (standard passenger car), 1936–1940. Horch's medium all-terrain standard car was equipped with all-wheel drive, two locking differentials, and independent suspension on upper and lower A-arms. To prevent high-centering, the EFm version had side-mounted auxiliary wheels. These were deleted from its successor, the Type 40 (1940–1943). Body by Trutz of Coburg. V8 gasoline engine, 3515 cc, 82 hp (60 kW).

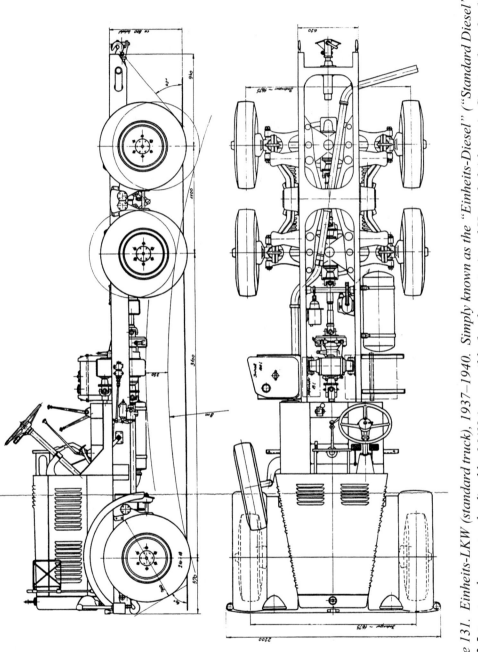

Figure 131. Einheits-LKW (standard truck), 1937–1940. Simply known as the "Einheits-Diesel" ("Standard Diesel"), this 2.5-ton truck was standardized by MAN (responsible for the engine) and Henschel (chassis). Six simplex wheels, independent suspension, six-wheel drive, and four differentials provided outstanding off-road capability. Six-cylinder diesel engine, 6234 cc, 80 hp (59 kW).

participating firms (MAN, Büssing-NAG, Magirus, Henschel, and Faun) had to use identical and therefore interchangeable components. The direct-injection diesel engine was developed by MAN; Henschel did the suspension. Because of all-wheel drive and tandem rear axles with single wheels, the three-axle Einheits-Diesel exhibited above-average off-road capability. However, only 7,100 were produced.

The Krupp-Protze (Figure 132) also had inline single rear wheels (i.e., non-duplex tires) and therefore a high level of off-road capability. This light truck, intended to tow light artillery pieces, incorporated the state of the art in chassis design, with nearly constant-camber independent suspension on upper and lower A-arms at the rear. Furthermore, it was powered by a highly advanced air-cooled boxer (horizontally opposed) engine, designed by Adolf Roth. The suspension of the Krupp-Protze, with six separate wheels, probably influenced the design of the Einheits-Diesel (standard diesel truck).

Figure 132. Krupp-Protze, 1934–1942. Although equipped with only rear-wheel drive, the Krupp-Protze also was capable of off-road service, because of its tandem rear axle and simplex tires. Its characteristic sloping hood was an added advantage, made possible by a horizontally opposed (boxer) engine. The designation "Protze" derives from the two-wheeled limber used by horse-drawn artillery and indicates its intended use as a tow vehicle for artillery pieces. However, in service, the Krupp was more often used as a light all-terrain truck. Four-cylinder horizontally opposed gasoline engine, 3308 cc, 54 hp (40 kW).

Halftracked tow vehicles with pneumatic tires on steerable front wheels and linked tracks at the rear were typical of World War II. Although basically similar designs, the U.S. halftracks had shorter tracks, as well as braked and powered front wheels; German halftracks had longer tracks and unbraked, undriven front wheels (Figure 133). By the end of the war, American

Figure 133. Halftrack tractor, 1934–1945. The German Army Ordnance Department classified halftrack tractors into six classes from 1 to 8 tons, referring not to their payload but to their towing capacity in medium terrain. For each class, a single firm was responsible for development to the point of production. The smallest class, shown here, was developed by DEMAG. To achieve adequate production numbers, Adler, Büssing-NAG, and Phänomen had to share the manufacture of the DEMAG D6 and, as of 1939, the D7. D7: six-cylinder gasoline engine (Maybach), 4170 cc, 100 hp (74 kW).

plants delivered 41,170 halftracks to U.S. and Allied forces, while German factories managed to produce approximately 48,000 vehicles.

Although closely associated with World War II, the halftrack is no war child. Beginning in 1901, Alvin O. Lombard of Waterville, Maine, built steam-powered halftracks fitted with skis or wheels at the front for the logging industry. In 1916–1917, he sold 104 examples, now gasoline powered, to the Russian army. Another American, Benjamin Holt, also delivered halftrack tow vehicles to the British during World War I. Among the best known of this genre were the Citroën Kégresse halftracks. Adolphe Kégresse was in charge of Tsar Nicholas II's motor pool and had transformed regular production Packard and Rolls-Royce cars into tracked vehicles for winter service. Fleeing the Bolsheviks, he landed in Paris, where he was able to win André Citroën over to his traction system. Using Citroëns converted to tracked drive, expeditions crossed the Sahara and Africa in 1923 and 1924–1925, and Asia in 1931–1932 (Figure 134).

For various reasons, American car makers had little reason to build military vehicles prior to the outbreak of war in September 1939. Only after May 1940, when the success of German Blitzkrieg tactics made clear to everyone the danger of a Nazi-ruled Europe, did the War Production Office issue a request for bids for a General Purpose Vehicle. In a phenomenally short

Figure 134. Citroën-Kégresse Autochenille, 1931–1932. Citroën conducted its Croisière Jaune, motorized expeditions through Asia, using Citroën halftracks designed by Kégresse-Hinstin. The China Group, traveling from east to west from Tientsin, consisted of seven six-cylinder vehicles; the Pamir Group, traveling eastward from Beirut, used seven four-cylinder halftracks (seen here). All were equipped with front-mounted winches and filled, puncture-proof tires. Four-cylinder engine, 1628 cc, 30 hp (22 kW); or six-cylinder engine, 2442 cc, 42 hp (31 kW).

time, this resulted in the Jeep. Otherwise, as late as mid-1941, not much was done other than planning and preparation, although in April 1941, Chrysler was the first auto plant to begin producing tanks.

The United States was not yet officially at war. (Germany did not declare war on the United States until four days after the Japanese attack on Pearl Harbor, which occurred on December 7, 1941.) However, the United States supported British North Atlantic convoys and delivered war material according to the Lend-Lease Act of March 11, 1941. The Lend-Lease Act allowed all opponents of the Axis Powers immediate access, without payment, to war material and supplies produced by the booming American armaments industry.

Within the framework of the Lend-Lease Act, Great Britain alone received 98,207 Army trucks, predominantly Chevrolet, GMC (Figure 135), Ford, Mack, and White. The British themselves produced trucks under the names Bedford (Figure 136), Ford UK, Austin, AEC, and others.

Figure 135. GMC 2.5-ton 6×6, 1943. During World War II, the General Motors Truck and Coach Division grew to be the largest American manufacturer of military vehicles. Among other products, GMC built more than 560,000 examples of the 2.5-ton CCKW 352 three-axle, all-wheel drive truck. After the war, some were fitted with producer gas generators, because gasoline and diesel fuel were almost unavailable in Europe. Shown here is a CCKW fitted with an Imbert generator. Six-cylinder engine, 4425 cc, 104 hp (approximately 77 kW) on gasoline.

Figure 136. Bedford QLD 4×4, 1943. Bedford, a subsidiary of General Motors' Vauxhall, produced slightly fewer than 250,000 military vehicles. The Model Q, Bedford's first forward-control truck, proved to be the most successful four-wheel drive truck in the British Army, with a payload of 3 tons. Six-cylinder engine, 3519 cc, 86 hp (approximately 63 kW).

The rush of events leading up to World War II produced a new type of vehicle: the Jeep. In the space of only five days, independent engineer Karl Knight Probst designed a multi-purpose vehicle meeting the specifications of the U.S. Army. He submitted his drawings to the War Production Office by way of the American Bantam Car Co. Bantam had to deliver 70 proto-type units in 49 days, a feat that could be accomplished only by purchasing entire assemblies from subcontractors. Continental supplied the engines; Studebaker delivered axles and brakes; Warner made the transmissions; and Spicer made the differentials. Bantam itself, licensee and manufacturer of an Americanized Austin Seven (Figure 78), contributed little more than the gauges and horn buttons.

On September 23, 1940, the first General Purpose Vehicle (the identifier GP soon entered the vocabulary of military slang as "Jeep") took to the Army Proving Grounds in Aberdeen, Maryland. The QMC, the Quartermaster Corps, beat the test vehicle without mercy, under the suspicious eyes of Willys-Overland and Ford engineers. Both firms were eager to land a govern-ment contract but were unable to meet the tight deadlines. After Bantam received a follow-on order for another 1,500 units, Willys and Ford presented their own general purpose military vehicles in November 1940, basically copies of the Bantam. The QMC selected the Willys-Overland as its standard vehicle. By the end of the war, Willys produced 361,349 of its MB model; Ford produced a further 277,897 GPW Jeeps. Bantam itself managed a mere 2,643 units, most of which were shipped to the Soviet Union. There and in Japan, they served as patterns for those nations' own military vehicles (GAZ, Toyota).

With its combination of all-wheel drive and the off-road capability of a military vehicle on the one hand, and the dimensions, operating costs, and handling of a passenger car on the other, the Jeep (Figure 137) was not only novel but also set the pattern for de-velopment of light, all-wheel drive military vehicles worldwide. Probst's design was

Figure 137. Quarter-ton 4×4 truck (Jeep), 1943. Familiar as the Jeep, the Willys-Overland Model MB and Ford Model GPW differed little in layout from the Bantam Model BRC-40. Whereas Bantam had planned on using a Continental engine (1830 cc, 45 hp), Willys and Ford chose instead an antiquated Willys engine. Both the Willys and Continental engines were side-valve units. Four-cylinder engine, 2199 cc, 54 hp (approximately 40 kW).

superior to Porsche's Kübel- and Schwimmwagen. It was considered a vital military asset that helped the Allies win the war and probably was the most significant vehicle of the 1940s. Its effects may be felt in our own time. Although it had only rear-wheel drive, the Jeep-based Willys Jeepster (Figure 138), introduced in 1948, was the progenitor of recreational off-road vehicles today and the current generation of sport utility vehicles (SUVs).

Figure 138. Jeepster, 1948. Jeep production continued after World War II. With only minimal changes, Willys created a sporty utility vehicle, the Jeepster. With a lengthened wheelbase, chassis lowering, Jeep-like bodywork, and more cheerful colors, the "Sports Phaeton" (as it was titled in the brochure) represented the starting point of an entire market segment of comfortable, angular-looking off-road vehicles: the 1962 Jeep Wagoneer series, joined by Ford and Chevrolet in 1966, and Range Rover in 1970. Four-cylinder engine, 2199 cc, 63 hp (approximately 46 kW).

The Mass-Produced Automobile
19?? to 19??

The Automobile Industry and Automotive Technology in the 1950s

In the immediate postwar years, the automotive industry of the Western nations experienced phases of rebuilding and of satisfying pent-up demand. The two decades between 1960 and 1980 are characterized by the weeding out of competing firms, the rise of the Japanese auto industry, and the aftereffects of two so-called energy crises of 1973–1974 and 1978–1979 on the automotive industry and technology.

While the United States was able to convert its undamaged plants back to civilian vehicle production after World War II, European plants resumed production as a function of how badly their facilities were damaged. British industry recovered first; the Germans came last.

North America, which had grown richer and more self-confident as a result of the war, was able to afford even larger cars than ever in the past. Shattered and demoralized, Europe had to make do with small to mid-sized cars. The 1950s underscored the dichotomy in automotive design that traced its roots to the 1920s: the European school, with its economical vehicles and emphasis on technological progress and quality, and the American idiom with its "gas guzzlers" and ostentatious bodywork that were subjected to costly facelifts with every new model year. Instead of engineering improvements, such as much-needed suspension development, Americans gave styling their highest priority (Figure 139).

In terms of styling, General Motors distinguished itself above all of its competitors. Beginning in 1949, General Motors presented its model lineup, with its own views on research and technology, at the Waldorf-Astoria Hotel in New York. These soon developed into the Motorama exhibitions. Amid girls and glamour, General Motors "dream cars" were presented to the public to prepare them for the styling of future models. In this way, General Motors could sell new ideas (and new models) in a risk-free environment and avoid a debacle such as Ford's stylistically odd Edsel of 1957.

Europeans adopted certain American styling elements in toned-down form, such as wrap-around windshields to replace one-piece "panorama" windshields, and accentuated fender

Figure 139. Studebaker Champion/Commander, 1950. Designer Raymond Loewy wanted to redirect the wartime generation's fascination from the airplane to the automobile. Apparently, his inspiration for the large "spinner nose" and prominent fendertop trim was Lockheed's P-38 Lightning fighter plane. Loewy's design was a success; in 1950, Studebaker had the best sales year in its history. Ford of Cologne adopted the theme with a globe on the hood of its 1952 Taunus 12M. Six-cylinder engine, 2780 cc, 85 hp (63 kW) or 4025 cc, 102 hp (75 kW).

edges instead of tailfins (Figure 140). At the same time, an independent European body style developed, led by the Italian *carrozzieri*. It was characterized by a shape free of ostentation, harmonious proportions, and good space utilization (Figure 141). Pininfarina, Ghia, Bertone, Michelotti, and Frua supplied designs to French, Italian, British, and German car makers. Individual national styles dating back to the 1930s began to disappear.

In the 1950s, German body manufacturers such as Autenrieth, Baur, Deutsch, Hebmüller, Karmann (Figure 142), Reutter, Rometsch, Rudy, and Wendler produced a number of quite attractive one-offs and small series, often with elements "borrowed" from the Italians. However, there was no single uniquely German body style until the NSU Ro 80 appeared in 1967. As in other countries, German coachbuilders, quite capable in the 1930s, were placed in a difficult position as the industry transitioned from body-on-frame construction to unit bodies. Special bodywork required large investments of capital and labor, which fewer and fewer customers were prepared to support. Of the roughly 400 coachbuilders and wagon makers that had existed in Germany since 1880, by around 1960, only 7 independent firms survived. Today, 40 years later, Karmann is Germany's only significant manufacturer of passenger car bodies. The situation in other countries is no different.

Figure 140. Cadillac Eldorado, 1959. The "Lightning" fighter-bomber must have had an enormous influence on General Motors' chief designer Harley Earl. Its twin-tail design, with small rudders, inspired Earl's penchant for automotive tailfins. These first appeared, quite discreetly, on the 1948 Cadillac, and grew from year to year, to achieve near-ridiculous proportions in 1959. V8 engine, 6391 cc, 345 hp (approximately 254 kW).

Figure 141. Fiat 1800/2100/2300, 1959–1968. The trapezoidal shape refined by Pininfarina entered the production car scene with the Fiat 1800 (shown here). It was followed by similar passenger car bodies for British Motor Corporation (BMC) in Britain and Peugeot in France, as well as variations by other car makers. A curved instead of single-piece "panorama" windshield, clearly defined shapes with minimal ornamentation, and clean, almost sober styling paid little heed to the lessons of aero-dynamics. Six-cylinder engine, 1795 cc, 85 hp (63 kW).

Figure 142. Volkswagen Karmann-Ghia, 1955. The coachbuilding firm that eventually became Karmann was founded in 1874. In 1902, Karmann built its first automobile body, in 1949 the first Volkswagen Beetle convertible, and, as of 1955, a coupe developed by Carrozzeria Ghia of Turin, followed in 1957 by a convertible version. Although the added cost over a Beetle, whose technology they shared, amounted to 2195 Deutsche Marks for the coupe and 2895 Deutsche Marks for the convertible (1957), by 1973 Karmann had produced a total of 445,300 examples. Four-cylinder engine, 1192 cc, 30 hp (22 kW).

In the early 1950s, as American auto makers embarked on a horsepower race that would lead to 240-hp middle-class sedans by 1959, most Europeans were happy simply to be able to afford some sort of enclosed vehicle. As after World War I, World War II was followed by countless cycle cars, minimobiles (Figure 143), and microcars. With these, former aircraft manufacturers, body and motorcycle makers, producers of electrical equipment and machine tools, and private individuals sought a piece of the action as Europe remotorized.

Most makers of microcars and minimobiles failed to notice that their development, production, and marketing costs represented a disproportionately higher fraction of their products' realizable retail prices than did the corresponding overhead costs of conventional small cars. Although these minimalist cars were less expensive, their makers had far smaller profit margins. With increased economic well-being, a desire for "real" cars, and the tightly calculated pricing of the Volkswagen Beetle, Citroën 2 CV, and Morris (Figure 144), the market for microcars and minimobiles was essentially doomed.

Figure 143. ISO Isetta, 1953–1955. In 1953, the Milanese scooter and motorcycle manufacturer ISO introduced an original concept to the market. In a remarkable departure from the shape and layout of a conventional car, the two-seater had a short, motorcycle-like wheelbase of only 150 cm (less than 5 ft), a narrow rear track eliminating the need for a differential, and a front door, soon copied by countless microcar manufacturers. ISO licensed the car to Germany (BMW), France (VELAM), and England (Isetta of Great Britain). Single-cylinder two-piston two-stroke engine, 236 cc, approximately 10 hp (7 kW).

In the 1950s and 1960s, other European manufacturers were able to join Volkswagen in exceeding the magic million-unit barrier with lower-middle-class products. These included the Fiat 600 (1955–1970), the Renault 4 CV (1947–1961), the Morris Minor (1948–1974), and the Mini (1959–2000, Figure 156). Automotive history was not written by prestigious marques such as Rolls-Royce, Cadillac, or Ferrari, but rather by ordinary mass-produced automobiles that gave mobility to millions and empowered them with entirely new lifestyles.

These million-sellers make it quite apparent that the engineering details buried underneath the bodywork are of no interest to customers. They accept the standard layout of an upright front-mounted engine and rear-wheel drive (e.g., Ford Model T, Austin Seven, and Fiat 1100) as readily as air-cooled (e.g., Volkswagen) or water-cooled (e.g., Fiat 600 and Renault 4 CV) rear-mounted engines, or front-drivers such as the Citroën 2 CV or Mini. In effect, customers have granted manufacturers complete design freedom and the ability to optimize production costs to the best of their ability. The only conditions are that the resulting technology must remain functional in everyday service, maintenance costs must be low, and customer service and spare parts must be available at all times. If these conditions are met, even antiquated

Figure 144. Morris Minor, 1948–1974. The Minor was Morris' first postwar car, still using a side valve engine but in a unit body. Over the years, it was given more powerful OHV engines and in 1961 became the first British automobile to reach a production run of one million units. After production ceased in Britain in 1971, foreign subsidiaries assembled CKD Minors ("completely knocked down") until 1974. A total of 1,293,331 units were built, with various bodies. Four-cylinder engine, 803 cc, 30 hp (22 kW) or 948 cc, 37 hp (27 kW) (OHV).

engineering concepts can remain successful well beyond their time. This has been proven time and again by the Benz Velo, the Curved Dash Oldsmobile, the Ford Model T, and the Volkswagen Beetle.

In the case of automobiles with unfamiliar styling and technology that is too far advanced, the approval of a certain minority is assured, but high production numbers are unlikely. Repeatedly, designers and auto makers have overestimated the public's willingness to identify with the unfamiliar, as typified by Rumpler's Tropfen-Auto (Figure 89), the Chrysler Airflow (Figure 106), or even the rather modest advances represented by Borgward's passenger cars (Figure 108). The success of the Citroën DS line is the exception that proves the rule.

The futuristic appearance of the Citroën DS (Figure 145) represented "a simply breathtaking, bold concept which anticipates the automotive technology of tomorrow—and beyond" (*Auto Modelle 1960/61*, p. 90). The DS line combined avant-garde body styling with a long wheelbase, front-wheel drive, and front disc brakes. A central hydraulic system provided power boost for brakes, clutch, transmission, and steering. Coupled to spring and damper units at the wheels, it provided hydropneumatic suspension. Sensors automatically leveled the vehicle, and the spring rates adjusted to the vehicle load. In addition, the ride height could be set manually, within a ground clearance range of 9 to 28 cm (3.5 to 11 in.). During its 20-year production run, Citroën delivered 1,454,705 examples.

Figure 145. Citroën DS, 1955–1975. Similar to the Citroën 7 CV Traction Avant (Figure 99), the DS model line also employed front-wheel drive in a long wheelbase for above-average roadholding. Contrary to expectations of the time, hydropneumatic suspension, pioneered by the DS, was not widely adopted by car makers, although Citroën continues to use a more developed form of the concept to this day. Production run was 1,454,705 units. Four-cylinder engine, 1911 cc, 75 hp (55 kW) (1960).

Despite impressive production numbers for such an individualistic car, it should be mentioned that Citroën was unable to survive as an independent company. Even in the early 1930s, the company, influenced by its chief engineer André Lefèbvre, embarked on a course driven by its "engineering philosophy" at the expense of purely commercial considerations. It survived only because of a takeover by the Michelin Group in early 1935. In 1974, Citroën passed into the hands of Peugeot SA.

The examples presented by Peugeot and Daimler-Benz show that even firms that once set the pace of engineering progress eventually adopted more cautious product policies. In this way, both firms were able to secure their own market positions.

This in turn reflects the change in the automobile's significance in society, a change that manifested itself in the late 1950s. The automobile was transformed from a technical product into a consumer good. No longer was the engineering capability of a company the sole determinant of the market success of a product. Now, modern management was required as well. Lawyers and economists took their places on auto makers' management boards. Engineering concepts and elegant detail design lost to corporate policies that placed market research ahead of engineers' dreams.

Opel, Vauxhall, and Ford Europe were proponents of this new corporate line. Even before World War II, these firms had already introduced the attitudes of their American corporate parents to the Old World—the philosophy of "good enough" rather than the old European ethic of "as good as possible." Firms based solely on engineering ability did not survive the 1960s. They failed, such as Borgward or Bugatti, or were absorbed, as in the case of Panhard, Citroën, DKW/Auto Union, and Lancia.

The formation of the European Economic Community (the Common Market) on January 1, 1958, led to a sudden increase in inter-European trade. It also resulted in tougher competition among individual producers. The second half of the 1960s, in particular, saw examples of production halts, various forms of cooperation, and mergers. These continued in the latter half of the 1970s, in part under pressure from the success of Japanese exports. Of more than 34 independent auto manufacturers in the United States and the "classic" auto-producing nations of Europe in 1960, only about 10 market-dominating firms and a handful of smaller companies remained in 1980 (Table 7).

The German Auto Industry: Reconstruction and Consolidation

In Germany, the postwar years were marked by reconstruction of demolished factories and infrastructure, recovery of the automobile and supplier industries, the unexpected success of the Volkswagen, and a regained self-confidence expressed in remarkable engineering and sports achievements. The basis for this resurgence was laid in March 1948, with the founding of the Bank Deutscher Länder (the Bank of German States), currency reform on June 20, 1948, and the introduction of a social market economy, rescinding price controls and price freezes.

Beginning in April 1948, the Marshall Plan provided additional aid in the form of goods and favorable investment credits. German entrepreneurs and industrialists were able to install state-of-the-art machinery and modernize their production facilities. For the coming technology race, this put German firms in an even better position than England and France, victors of the recently ended war. In the second half of 1948, total industrial production rose by almost 50 percent. Truck production climbed 190 percent from 1948 to 1949, and passenger car output rose by an even higher margin of 350 percent.

In 1950, the truck and motorcycle plants in the Western Zone produced more units than the entire Reich in 1938. Passenger car production did likewise in 1951. All this occurred within the framework of a geographically and ideologically divided country, with fewer auto manufacturers than before the war. Stoewer, in Stettin, found itself in Polish territory. Hanomag concentrated on tractors and trucks, while Maybach and Adler withdrew entirely from auto manufacturing. BMW lost its plants in Eisenach, Auto Union likewise its facilities in Saxony, both behind the Iron Curtain. Opel not only lost its truck plant in Brandenburg, but also had to turn over all Kadett tooling and production facilities to the Soviet Union as reparations. From 1947 to 1956, the Soviets continued to build the Kadett as the "Moskvitch" (Son of Moscow). As a result, a possible Volkswagen competitor was eliminated from the outset.

TABLE 7
COMPANIES AND GROUPS—1960, 1980, AND 2000

		1960	1980	2000
D		Auto Union/DKW		
		BMW	BMW including Glas	BMW including Mini, Rolls-Royce (as of 2003)
		Borgward, Goliath, Lloyd		
		FMR/Messerschmitt		
		Ford Germany	Ford Germany	Ford Germany
		Glas		
		Mercedes-Benz	Mercedes-Benz	DaimlerChrysler including Smart & Chrysler (see USA)
		NSU		
		NSU Fiat		
		Opel (GM)	Opel (GM)	Opel (GM)
		Porsche	Porsche	Porsche
		Trabant (East Germany)	Trabant (East Germany)	
		Volkswagen	Volkswagen including Audi/NSU	Volkswagen including Auto Union, Audi/Lamborghini, Seat, Skoda, Bugatti, Bentley, and Rolls-Royce (until 2002)
		Wartburg (East Germany)	Wartburg (East Germany)	
F		Alpine		
		Citroën		
		Facel-Vega		
		Grégoire		
		Panhard		
		Peugeot	PSA including Peugeot, Citroën, and Chrysler EU: Matra, Simca, Talbot, Sunbeam	PSA (Peugeot, Citroën)
		Renault	Renault including Alpine and AMC (see USA)	Renault including Dacia and Nissan (see J)
		Simca		
		Vespa		
GB		AC, Allard, Alvis, Amrstrong-Siddeley	AC	AC
		Bristol, Berkeley	Bristol	Bristol, Carbodies, Caterham
		David Brown including Aston Martin/Lagonda	Aston Martin Lagonda	
		BMC including Austin (-Healey), MG, Morris, Riley, Wolseley, and Vanden Plas	BL Limited including Austin, Carbodies, Jaguar/Daimler, MG, Mini, Morris, Princess, Rover, Triumph, Vanden Plas	
		Ford GB	Ford GB	Ford GB
		Jaguar/British Daimler		
		Jensen, Lotus, Morgan	Lotus, Morgan, Panther Reliant	Ginetta, Jensen, Lea Francis, Lynx, Marcos, Morgan, Reliant
		Rolls-Royce/Bentley	Rolls-Royce/Bentley	
		Rootes Group including Hillman, Humber, Sunbeam and Singer		
		Rover		Rover, including MG
		Standard/Triumph		
		TVR	TVR	TVR
		Vauxhall (GM)	Vauxhall (GM)	Vauxhall (GM)

161

TABLE 7
COMPANIES AND GROUPS—1960, 1980, AND 2000
(CONT.)

	1960	1980	2000
USA	American Motors Corp. including Rambler and Metropolitan	AMC including Jeep and Eagle (see F/Renault)	
		Avanti	
	Checker	Checker	
	Chrysler Corp. including Chrysler, De Soto, Dodge, Imperial, and Plymouth	Chrysler Corp. including Chrysler, Dodge, and Plymouth	DaimlerChrysler including Chrysler, Dodge, Plymouth, Jeep, and Mercedes (see D)
		Excalibur	
	Ford Motor Co. including Ford, Lincoln, and Mercury	Ford Motor Co. including Ford, Lincoln, and Mercury	Ford Motor Co. including Ford, Mercury, and Premier Group (Aston Martin, Jaguar, Lincoln, Volvo, and Land Rover)
	GM including Buick, Cadillac, Chevrolet, Oldsmobile, and Pontiac	GM including Buick, Cadillac, Chevrolet, Oldsmobile, and Pontiac	GM including Buick, Cadillac, Chevrolet, Geo, Hummer, Lotus, Oldsmobile, Pontiac, Saab, Saturn and Fiat (20%)
	Studebaker Willys		
I	Abarth Alfa Romeo Autobianchi Ferrari Fiat Lancia Ferrari Maserati DeTomaso	Alfa Romeo Ferrari Fiat including Abarth, Autobianchi, and Lancia Lamborghini Maserati DeTomaso	Fiat (see GM/USA) including Alfa Romeo, Ferrari, Lancia and Maserati
J	Daihatsu Datsun/Nissan Isuzu Mazda Mitsubishi Prince Subaru Suzuki Toyopet/Toyota	Honda Nissan/Datsun Isuzu (GM) Mazda Mitsubishi Subaru Suzuki Toyota including Hino and Daihatsu	Honda including Acura Nissan including Infiniti (see F/Renault) Isuzu (GM) Mazda (Ford) Mitsubishi (DaimlerChrysler, Volvo) Subaru (GM) Suzuki Toyota including Hino, Daihatsu, and Lexus
KO		Daewoo Hyundai Kia including Asia Ssang Yong	Daewoo including SSang Yong Hyundai incl. Kia and Asia Samsung

The Volkswagen plant began production of civilian vehicles in 1945. Daimler-Benz followed in 1946, Opel in 1947, Ford in 1948, and Borgward in 1949. Auto Union, newly regrouped in the West with the DKW brand, plus Goliath, Lloyd, Gutbrod, and Porsche all resumed German production in 1950. By the end of the reconstruction phase, 1952, BMW of Munich was the last of the major auto manufacturers to resume operations.

Just as after World War I, as a result of years of political isolation, Germany had lost touch with world-class technology. The technical advantage enjoyed by German industry in several key areas in the 1930s had dissolved by the late 1940s. German auto manufacturers and suppliers had to shop abroad for loans and licenses.

In shaping its 1949 Hansa 1500 (Figure 146), Germany's first new postwar design, Borgward drew heavily on the American Kaiser-Frazer of 1947 (Figure 147). Teves offered the German auto industry (initially with little success) valve rotators ("Rotocaps"), hydraulic lifters, and power brake boosters licensed from American firms. Beginning in 1950, Borgward attempted to introduce an automatic transmission, licensed from Hobbs of England, to the European middle class.

Figure 146. Borgward Hansa 1500, 1949–1952. While other European car makers resumed production of their prewar models, more or less unchanged, Borgward introduced an all-new model with its Hansa 1500. After the Hanomag Kommissbrot ("Breadloaf Car," which was shaped similarly to a loaf of commissary bread) of 1924, it was the first German production car fitted with pontoon bodywork. Its shape was inspired by the early postwar Kaiser-Frazer models in the United States (Figure 147). Four-cylinder engine, 1498 cc, 48 hp (35 kW).

Figure 147. Kaiser-Frazer, 1947–1950. After the war, the Kaiser-Frazer Corporation introduced a completely new car. Its pontoon body, designed by Howard "Dutch" Darrin, captured the public's attention with its understated, harmonic lines and lack of chrome excesses. Singer of Britain (1948) and Borgward of Germany (Figure 146) copied the shape of the twin Kaiser and Frazer models but were unable to match the quality of the American originals. Six-cylinder engine, 3707 cc, 110 hp (81 kW).

Founded in 1949, the Federal Republic of Germany went through a consolidation phase between 1952 and 1958, marked by full employment and equal status for the young nation in international economic organizations. The auto industry enjoyed a period of vigorous growth, thanks to cautious fiscal and monetary policy, avoidance of strikes, and restrained wage demands on the part of unions. In 1954, the size of the German automotive industry, including trucks and buses, exceeded that of France; in 1956, it surpassed Britain. Steady prices and rising demand on worldwide markets led to increasing exports. In 1954, the Federal Republic of Germany exported more cars than France or the United States, and in 1956 exceeded British exports. For almost 20 years, Germany remained the top exporter and assumed second place in production behind the United States.

Because of funds invested in reconstruction, technical improvements to vehicles were limited in scope. In areas such as radial tires (Citroën 2 CV, 1949), disc brakes (Citroën DS 19, 1955), suspensions (Earle S. McPherson of Ford/USA, introduced by Ford GB in 1951), electric fuel pumps, fuel economy, and ride comfort, German manufacturers fell behind their French and British competition. However, the Germans gained an advantage in the field of gasoline fuel injection. After the Allies prohibited construction of aircraft and aircraft engines after the war, Bosch, in cooperation with Gutbrod and Goliath, transferred its expertise in (direct cylinder) gasoline fuel injection to passenger car two-stroke engines. Fuel injection is especially suited for port-controlled two-strokes because it eliminates scavenging losses, reduces consumption and emissions, and increases power output. At the 1951 International Automobile Exposition (Internationale Automobil-Ausstellung, or IAA), Gutbrod exhibited a fuel-injected engine. Goliath displayed a complete car, a sports coupe powered by an injected engine (Figure 148). After decades of experimentation by many firms, (direct) fuel injection was finally available, in cars affordable for every man, as both Gutbrod and Goliath commenced production of their injected two-strokes in 1952.

Because fuel injection also results in higher specific output in four-stroke engines, injected Offenhauser/Meyer-Drake engines first ran at the 1949 Indianapolis 500. The injection system provided by Stuart Hilborn for the Howard Keck Special employed continuous injection

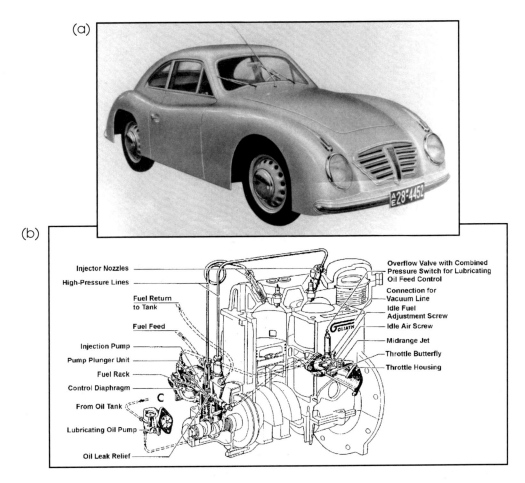

Figure 148. Goliath GP 700 E Sport, 1951. Goliath employed a unique sports coupe (a) as a test bed and promotional device for fuel-injected two-stroke engines (b). Rudy of Delmenhorst built two examples, while Rometsch of Berlin made 25. The injected engines installed in Goliath's later passenger cars and delivery and off-road vehicles proved to be quite reliable. Two-cylinder two-stroke engine, 688 cc, 29 hp (21 kW).

into the intake manifold. These systems could not provide reliable mixture over a wide engine speed range and therefore were not yet suitable for passenger car use.

In Germany, fuel injection development also first concentrated on sports and racing cars. The participating firms—Bosch, Mercedes, and Borgward—chose direct cylinder injection, familiar from aircraft and two-stroke engine practices. Beginning in August 1954, Mercedes equipped its mass-produced 300 SL with direct injection as standard equipment. Identical systems were installed on the W 196 Formula 1 race car of 1954, and the 3-liter 300 SLR sports racer of 1955 (Figure 149).

Figure 149. Mercedes-Benz 300 SL, 1954–1963. The 300 SL coupe was derived from a sports racer that first entered competition in 1952. With a space-frame chassis, power brakes, gasoline fuel injection, and distinctive gullwing doors, it was one of the most technologically interesting sports cars of its day. The Bosch injection pump, injecting directly into the combustion chambers, was mounted below the engine, which was canted at 45 degrees to lower the hoodline. Six-cylinder engine, 2996 cc, 215 hp (158 kW).

While victories by Mercedes' racing and sports racing cars reinforced the worldwide prestige achieved by Volkswagen's phenomenal export success, Borgward's direct-injected 1.5-liter sports racer, first raced in 1954, achieved its zenith in the 1959 Cooper-Borgward. Unfortunately, Borgward chose not to install fuel injection in its production cars. Mercedes, on the other hand, put injected engines in its top-of-the-line 300 S (beginning in 1955) and 300 d (from 1957) and, as of 1958, in its 220 series (Figure 150). For cost reasons, Mercedes switched from direct injection to manifold injection in 1957.

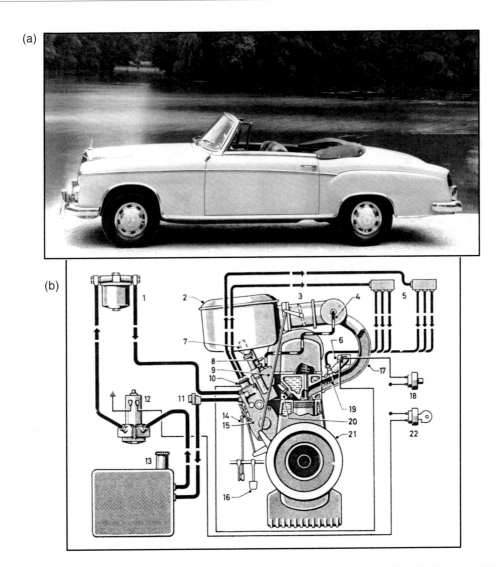

Figure 150. Mercedes-Benz 220 SE coupe and cabriolet, 1958–1960. In 1955, Mercedes introduced fuel injection to mass-produced automobiles, initially direct injection derived from sports cars, and as of 1957 indirect injection (b), first offered on the 220 SE (a). The system employed a Bosch two-plunger pump (item 15 in the schematic), which supplied fuel to two fuel manifolds 5 and from there to the injectors 19 mounted in the intake manifold runners 17. The performance gain compared to the carbureted 220 S model amounted to 15 hp (11 kW). Six-cylinder engine, 2195 cc, 115 hp (85 kW).

Schematic of manifold fuel injection with control devices and fuel flow: 1 *Main fuel filter;* 2 *Air filter;* 3 *Throttle housing;* 4 *Idle auxiliary air supply;* 5 *Fuel manifolds;* 6 *Thermo switch for cold-start solenoid;* 7 *Intake air temperature sensor;* 8 *Coolant temperature sensor;* 9 *Barometric pressure sensor;* 10 *Start solenoid;* 11 *Regulator housing;* 12 *Fuel feed pump;* 13 *Fuel tank;* 14 *Fuel rack;* 15 *Injection pump;* 16 *Throttle pedal;* 17 *Intake manifold;* 18 *Starter pushbutton;* 19 *Injector nozzle;* 20 *Spark plug;* 21 *Engine;* 22 *Ignition switch.*

From 1952, most German auto makers took part in national and international motorsports events. Countless wins by Porsche but, above all, victories by the Mercedes 300 SL (Figure 149), which won the 1954 and 1955 World Championships and thereby continued the string of prewar successes chalked up by the "Silver Arrows," expressed the exuberance of postwar Germany—and the regained self-confidence of its people. Firms that were almost unrecognized internationally made their marks on the world record scene, usually dominated by Anglo-American efforts— even if they competed only at the bottom end of the engine displacement scale. In 1951, a 700-cc Goliath, still fitted with a carbureted two-stroke, established 38 world records. In 1954, its feats were repeated by a 350-cc Lloyd. A year later, a 192-cc three-wheeled bubble car—a *Kabinenroller*—by FMR/Messerschmitt broke 25 existing records. A record of a different sort was established in 1953, when two drivers thrashed their standard production 700-cc Goliath sedan through the Australian outback, from Melbourne to Perth by way of Adelaide, achieving an average speed of 80.6 km/h (50 mph) and 37 mpg.

These and other successes had an enduring effect on the public's perception of the German auto industry. By the end of the 1950s, Germany had achieved a position of technological and economic leadership.

The Beetle and Mini Establish New Directions

Unlike other German firms, the Volkswagenwerk did not take part in motorsports. Instead, it devoted its attention to developing a car that a commission appointed by Britain's Society of Motor Manufacturers and Traders had given little chance of success. According to the report of the commission, "The vehicle does not meet the fundamental technical requirements of a motorcar. As regards performance and design, it is quite unattractive to the average motorcar buyer. It is too ugly and too noisy...a type of car like this will remain popular for two or three years, if that. To build the car commercially would be a completely uneconomic enterprise..." (Nelson, Walter Henry, *Small Wonder*, Little, Brown & Co., Boston, 1965/1970). Further, the report predicted that if by some chance the car should go into production, "it would mean no undue economic competition on the world market against British products." After this report was issued, the Control Commission for Germany (the British military government) appointed Heinrich Nordhoff as director of the Volkswagenwerk, effective January 1, 1948.

Despite the heavy political baggage of both the firm and indeed Germany, Nordhoff, who had managed the Opel truck plant in Brandenburg, turned what was admittedly a technically backward automobile into the most successful automobile maker in history. He achieved this by means of effective customer service, quality improvements, and model development such as had never been seen. Every year, immediately after the August plant holiday, an improved, more attractive, and often more affordable Volkswagen appeared on the scene. It was not Porsche's design and layout, but rather Nordhoff's detail improvements that permitted the Volkswagen to achieve its phenomenal success (Figure 151).

Figure 151. Volkswagen Type 1 (Beetle), 1946 to present (Mexico). During a production life that continues today, the Beetle experienced fewer external changes than did the Ford Model T in its 19-year run. Before German production ended in 1978, the Beetle underwent a process of continuous development that made it a highly desirable product. Shown here is a 1966–1967 model, one of more than 21 million examples produced, available as either the standard 1300 A or as a 1300/1500 equipped with a 1.5-liter 44-hp engine. Four-cylinder rear-mounted horizontally opposed ("boxer") engine, 1285 cc, 40 hp (29 kW).

Until early 1950, the Volkswagen had no competitors on the German market. As Goliath, DKW, and the Fiat 1100 appeared on the scene, they proved to be disappointments because of their high prices and service, which did not begin to approach the Volkswagen levels of efficiency that customers had come to expect. In the postwar years, mini- and microcars (Figure 152) sprang up like mushrooms after a rainstorm, but Volkswagen's price reductions, down to 3790 Deutsche Marks (1955–1961) for the standard Volkswagen, soon strangled most of these pretenders to the economy car throne. Often employing high technology, these companies sought to fill the gap between motorcycles and the Volkswagen. However, by around 1960, Volkswagen's challengers had disappeared from the scene.

Volkswagen, on the other hand, announced one success after another. In 1950, its market share in Germany stood at 41.5 percent. Opel had 21.6 percent market share, compared to 36.8 percent in 1938. Auto Union/DKW held only 1.1 percent of the market, whereas in 1938, it had 17.88 percent. In 1953, Volkswagen's Brazilian plant came on-line, and in 1955 the one millionth Volkswagen rolled off the assembly line. Meanwhile, the export market had grown to encompass more than 100 foreign countries, with the United States as the most important Volkswagen market.

On American shores, the Volkswagen was known as the Beetle, a designation soon adopted by the Germans. In the United States, the Beetle launched an "automotive counterculture."

Figure 152. Zündapp Janus, 1957–1958. Although they sometimes showcased unusual technology, microcars often were ridiculed unjustly. The Janus, developed by Dornier, was one of few passenger cars (along with the American buggies around 1905) with a true mid-engine layout. The Janus was economical to manufacture because of its almost identical side panels and its interchangeable front and rear doors. Its unit body featured back-to-back seating and above-average performance. Single-cylinder two-stroke engine, 248 cc, 14 hp (10 kW).

As noted by the caption accompanying a display of a 1949 Volkswagen in the Henry Ford Museum in Dearborn, Michigan, "The car was a clear-cut alternative to Detroit's stylistic excesses. Fun to drive, cheap to operate, and superbly constructed, the VW had become an inverse status symbol. The car ultimately exerted a tremendous influence in the American driving public and on the Detroit automobile industry."

Buoyed by the outcome of the war, American self-confidence led to incredibly large automobiles in the 1950s and 1960s, to gigantic engines and sheet-metal excesses with tailfins, cascades of chrome, and candy colors. "Dream cars" resembling spaceships, and names such as Golden Rocket, Star Chief, Starfire (Figure 153), Comet, Skyliner, and Galaxie, promised new and loftier aspirations once the superpower had subjugated the earth itself.

Instead of boarding their own personal spacecraft, hardheaded, practical customers often chose economy cars, which, except for the Studebaker Lark and the Rambler American, were offered only by European makers. By early 1959, imports had captured 10 percent of the American market, led by Volkswagen despite Wolfsburg's rationing of cars to North America to deliver sufficient units to other markets. The "Volkswagen go home" sentiment expressed in *LIFE* magazine was taken in its intended spirit, as a joke. However, United Auto Workers

Figure 153. Oldsmobile Starfire, 1958. With designs such as Howard Darrin's Kaiser-Frazer (Figure 147), Raymond Loewy's 1953 Studebaker Commander Starliner/Starlight, and the 1963 Avanti (to name only a few), the American car industry showed it could create captivating designs. However, it was perhaps best known for its chrome machines. These included the 1958 Oldsmobiles, whose quadruple chrome side trim was likened by many satirists to musical notation. V8 engine, 6081 cc, 265 to 312 hp (approximately 195 to 230 kW).

president Walter Reuther was completely serious when he demanded that German cars should carry a special tariff to compensate for lower German wages. The "American Peril" of the 1920s, with 33 percent imports posing a real danger for German auto makers, had unexpectedly turned into a "German Peril" for the U.S. auto industry. For Americans who were accustomed to victory, this was a new experience and only a taste of what the Japanese would serve a few years later.

By the end of 1959, 400,000 Volkswagens were registered in the United States. The amazing sales success of Volkswagen was built entirely on economic grounds. The average American worker, earning $442 per month, needed 3.68 months to purchase a Volkswagen export model at $1,625. By comparison, a skilled industrial worker in Lower Saxony, with gross monthly earnings of 430 Deutsche Marks in 1956, had to work for 8.8 months for a standard Volkswagen at 3790 Deutsche Marks, and 11-1/2 months to afford a Volkswagen with export trim, priced at 4600 Deutsche Marks. This was only a modest improvement over the situation in Germany in the 1930s. In the Federal Republic, the relationship between earnings and car prices continued to pose a formidable obstacle to mass motorization.

However, let's return to the United States. A standard American car, in its lowest-priced, two-door version, cost $2,132 (1959 Ford Custom 300 business sedan) or $2,160 (1959 Chevrolet

Biscayne utility sedan). The buyer could hardly use the vast expanses of space inside these cars. On average, 85 percent of the time, American cars carried only 1.6 people, on short trips of less than 12 miles. For years, consumers had been bombarded with the idea that "Mr. Average" needed a big car, but this began to lose credibility.

Detroit reacted quickly. To stem the tide of imports, the "Big Three" introduced so-called "compact cars" in late 1959 (although these would still be considered large cars by European standards). Chrysler had a Euro-American hybrid, the Valiant; Ford a reduced-scale standard car (Falcon), and General Motors a larger variation on the Volkswagen theme, the Chevrolet Corvair (Figure 154). Similar to its inspiration, the Corvair was powered by an air-cooled horizontally opposed ("boxer") engine at the rear, at 176 kg (388 lb) considerably heavier than the Beetle powerplant, which tipped the scales at 110 kg (242.5 lb).

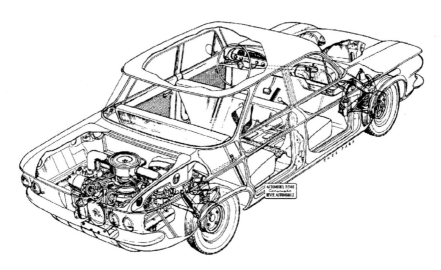

Figure 154. Chevrolet Corvair, 1960–1969. For a mass-produced automobile aimed at America's "defensive drivers," the Corvair featured a number of unconventional and daring engineering details. The engine location behind the rear axle, excessive engine weight of 176 kg (388 lb), its aluminum alloy horizontally opposed six-cylinder air-cooled engine, and its two-link semi-trailing rear suspension resulted in high manufacturing costs and handling characteristics that were no longer state of the art. Six-cylinder horizontally opposed engine, 2684 cc, 142 hp (104 kW).

The weak points of the Beetle, including a tendency toward oversteer (see Appendix), marginal straight-line stability, and crosswind sensitivity, were magnified by the more powerful Corvair and often overwhelmed the average American motorist, schooled as he was in "defensive driving." Dangerous handling characteristics and Corvair's typical rollover accidents prompted a heretofore unknown attorney by the name of Ralph Nader to unleash a nationwide safety campaign that severely tarnished the image of General Motors—and rubbed off on Volkswagen as well.

In the General Motors refusal to improve the active safety of Corvair by means of radial tires, thicker stabilizer bars, and different shock absorbers, Nader saw more than only an auto maker's corporate greed. He accused the auto industry of pervasive engineering blunders, and he claimed that auto firms ignored patents and discoveries that promised better vehicle safety, impeded long overdue improvements, and, with their growing power, showed increasing disregard for the health and well-being of their customers.

Nader's book, *Unsafe at Any Speed*, was published in 1965. At Chevrolet, Corvair sales plummeted from 235,528 units in 1965 to 15,399 in 1968. Volkswagen was able to increase its sales initially and did not show a decline until 1970, which occurred for other reasons.

Unintentionally, Chevrolet and Volkswagen had touched off a transformation in consumer consciousness. An ever more critical public was no longer prepared to accept antiquated design principles, technical defects, or manufacturing flaws. Auto makers suffered through the painful experiences of recall campaigns and product liability suits. In the process, expensive research and a high human cost again proved that a rear-engine design was not suitable for family cars capable of high speeds, because active safety of such vehicles did not represent the current state of the art. This had, after all, been recognized in the 1930s. If anything, Volkswagen may well have delayed the general introduction of safer family cars by one to two decades.

The Corvair was by no means the only imitator of the Beetle. Following the successful lead of Volkswagen, the rear-engined layout was copied by Fiat, Renault, Hillman, Simca, BMW, NSU, and several Soviet Bloc firms. All apparently regarded a rear engine as the first rung of a profit ladder leading to ever-climbing production numbers. Had they analyzed the situation more carefully, they would have realized that the success of Volkswagen came not from its layout but from rational manufacturing methods, higher utility, universal applicability, effective customer service, and everyday reliability—attributes that could just as well be realized with a standard layout or front-wheel drive.

DKW, Goliath/Hansa, Lloyd, Panhard, Citroën, and Saab, to name a few, were to some extent Beetle competitors but remained unswayed by the rear-engine layout of Porsche. Instead, they improved the reliability and efficiency of their front-drive cars. First among these was Citroën SA. The firm's 2 CV (Figure 155), the "French Volkswagen," was the first production car ever to be fitted with radial tires as standard equipment, as early as 1949. The United States and Britain stuck to the standard layout—with the exception of the Chevrolet Corvair and the twin Austin Seven/Morris Mini-Minor (Figure 156), usually simply called the Mini.

British engineers were often underestimated by their German and French colleagues. One expected them to build conservative prestige limousines and bone-shaking sports cars but never a mass-produced car with instant appeal. However, this is precisely what BMC's chief engineer, Alec Issigonis, designed in response to the 1956 Suez Crisis and its concomitant effects on the British economy, which prompted the public to buy smaller cars.

Figure 155. Citroën 2 CV, 1949-1990. Thanks to its roll-up top and rear lid, tubular steel seat frames, and lightweight design, Citroën's 1949 "Duck" weighed only 498 kg (1098 lb). Front-wheel drive, standard Michelin steel-belted radial tires, and linked front and rear springing on each side gave the 2 CV a comfortable ride—and spectacular body lean in cornering. More than five million were built. Shown here is its original 1949–1951 incarnation. Two-cylinder horizontally opposed engine, 375 cc, 9 hp (6.6 kW).

Figure 156. Mini, 1959-2000. In 1959, British Motors Corporation presented Alec Issigonis' "space achievement." Despite small exterior dimensions, the Mini offered an amazingly large interior. With a front-mounted transverse engine and front-wheel drive, the Mini is the ancestor of modern compact and midrange automobiles. Independent front and rear suspension was unusual at the time, at least for British products. Four-cylinder engine, 848 cc, 30 hp (22 kW) (1959).

The Mini, presented in 1959, featured a host of innovations. With the exception of limited-production sports cars by Alvis and BSA in the 1920s and 1930s, the Mini was the first British car with front-wheel drive, the first production car with a transverse four-cylinder inline engine, the first to have a transmission located below the crankshaft sharing the same crankcase, and the first to employ independent suspension by elastomeric elements. These were later replaced, in the fall of 1964, by hydrolastic suspension. Other design features that enabled the 3.05-meter (10-ft) long and 1.41-meter (4.6-ft) wide car to accommodate four occupants were its long wheelbase (2.04 meters [6.7 ft]), minuscule 10-inch wheels located far outboard at the corners with correspondingly small wheel wells, upright seating, and sliding instead of crank-operated side windows, which permitted additional "elbow room."

Momentous changes are often not perceived as such—at least, not at first. Therefore, the ever-conservative British public held back, regarded the Mini's design as much too clever to have everyday utility, and bought old familiar English iron instead. Eventually, word spread that the Mini embodied an ingenious compromise between price, size, interior space, and performance. In four and a half years, by early February 1965, the first million examples had been built. For comparison, the Beetle took ten years to reach the same mark.

Exports accounted for almost a third of production. The main customers were the Commonwealth and members of the European Free Trade Association (EFTA). Although the Mini also found customers within the European Economic Community, German buyers remained few and far between. Despite certain advantages over the Volkswagen, the Mini did not appeal to the taste or mentality of conservative German customers, and initially it was too expensive. The 1960 price was 5175 Deutsche Marks, compared to the Volkswagen Export priced at 4600 Deutsche Marks.

British Motors Corporation transferred the engineering details of the Mini—front-wheel drive, transverse engine ahead of the axle, and long wheelbase—to the larger 1100 model (as of 1962) and the 1800 Maxi (from 1964). The once stodgy maker of conventional cars had transformed itself into a "builder of an engineering trend-setter of the first order" (Daniels, Jeff, *British Leyland: The Truth About the Cars*, Osprey, London, 1980, p. 36). Despite its success, BMC neglected to implement a program of model development *à la* Volkswagen. The first warning signal came in the 1960s, when imports of the Mini to the United States were halted because they could not meet the latest safety standards. The manufacturer of the Mini ignored these signals, as it had ignored Continental European imitators that appeared in droves beginning in 1965, and offered buyers more. These cars included Volkswagen/Audi, with the Golf, Audi 50, and Polo models, which entered the scene in 1974.

The Mini was in production until 2000 and will be replaced by a new, slightly enlarged Mini in 2001, under BMW management. The Beetle continues to roll off the assembly line in Puebla, Mexico. More than five million examples of the Mini have been built; the Volkswagen Beetle tally is more than 22 million. In the 1960s, the Mini had an incredible career as a rally car. Ironically, the Beetle—conceived in the 1930s for a totalitarian, egalitarian, socialist society—

unleashed instead an outburst of four-wheeled fun that celebrated individuality. California's dune buggies, based on Volkswagen mechanicals, gave birth to a new automotive species—replicars, which were leisure vehicles fitted with bodywork resembling antique Bugattis or other nostalgic designs. American Formula Vee racers emigrated to Europe where, as of 1966, they mutated into that continent's own Formula Vee and Formula Super Vee race series.

Because of the obvious advantages of the Mini—low manufacturing costs, increased efficiency due to elimination of the right-angled ring-and-pinion final drive, and more interior space—virtually all modern compact and midsize cars emulate the Mini pattern. In the United States, even luxury cars including Lincoln and Cadillac are built using the Mini layout. Indeed, the mighty American auto industry has finally adopted European technology, reversing a trend established after the end of World War I.

Unlike the Mini, the Volkswagen Beetle had greater impact on economic history than on automotive history. As a symbol of the so-called "German Economic Miracle," it brought in billions in foreign exchange and secured the reputation of German quality throughout the world.

Mass Motorization, and American Legislative and Oil Crises

In the early 1960s, the negative effects of mass motorization became all too apparent, especially in and around American cities. The American legislature felt compelled to take action, even more so under the pressure of the 1970s energy crises. These actions reverberated throughout all auto-making nations, but one immediate effect was that the U.S. auto industry experienced its most serious sales slump in decades.

Initially, the Western industrialized nations enjoyed rising automobile sales, fed by pent-up demand from World War II, a steadily improving standard of living, and, in the Old World, the founding of the European Economic Community (EEC). As of January 1958, this ensured the six founding nations—France, Belgium, the Netherlands, Luxembourg, Italy, and West Germany—with a home market of 165 million inhabitants, without duties or tariffs.

In terms of production figures, Europe had been left in the dust by the United States since 1906 and thrown back by two world wars. However, in 1969, Europe produced slightly fewer than 9.44 million passenger cars, more than the United States and Canada combined (Table 8). The primary contributors to Europe's output were France and Germany, even as non-EEC member Great Britain increasingly lost significance as a car-making nation.

By 1980, the British auto industry had already been infiltrated by foreign firms or had gone bankrupt. By the mid-1980s, General Motors subsidiary Vauxhall and Ford UK imported half of the cars they sold in Great Britain, most from Germany. The Rootes Group had become part of Chrysler UK in the 1960s, which in turn was taken over by PSA/Peugeot in 1978. By 1980, BL Ltd. was the single remaining major auto maker in Britain but for years operated unprofitably. In 1987, it had to sell its truck division to DAF; in 1994, its car division went to BMW. Today, it produces transporters under the LDV label. By 1998, what had been the second-largest auto-producing nation in 1956 (behind the United States) had dropped to seventh place behind Japan, the United States, West Germany, France, Spain, and Canada (Table 9). Now that Rolls-Royce/Bentley, Jaguar, and Aston Martin are in foreign hands, Rover/MG remains the largest—indeed the only—wholly British-owned vehicle manufacturer.

TABLE 8
PASSENGER CAR AND STATION WAGON PRODUCTION
BY GLOBAL REGION, 1938 TO 1998

Year	North America USA and CDN	Western Europe* EEC, EU, and EFTA	Far East J, AUS, and IND	Soviet Bloc	Latin America	World Total
1938	2,124,746	868,409	8,500	40,475	–	3,043,831
1950	6,950,660	1,105,416	20,896	ca. 97,000	–	8,174,032
1960	7,028,390	5,079,348	434,191	271,821	121,733	12,935,583
1970	7,468,941	9,531,378	3,505,000	831,900	475,785	22,288,274
1979	9,464,733	10,344,354	6,726,318	2,251,857	1,464,018	31,540,390
1980	7,265,510	9,447,529	7,383,653	2,279,300	1,522,133	29,357,297
1990	7,000,598	13,013,659	10,744,311	1,729,899	1,140,925	35,405,868
1998	8,459,872	14,470,551	8,665,071	–	2,497,509	38,656,268

* 1979 (890,824) and 1980 (976,216) excluding Spain.
Source: *Automobil-Revue und Catalog*, various years. Figures comparable only under certain circumstances.

In the United States, neither the Ford Falcon/Mercury Comet, the Plymouth Valiant, nor even the Corvair (conceived as a "Beetle killer") were especially successful. All three firms pushed larger models, initially still labeled "compacts." By 1970, these were as large as ever; big cars bring big profits (Figure 157). Customers and the auto industry again had found their lowest common denominator.

TABLE 9
PASSENGER CAR AND STATION WAGON PRODUCTION
BY COUNTRY, 1938 TO 1998

Year	USA	D	F	GB	J	KO
1938	2,000,985	276,592	189,691	341,028	8,500	To 1984:
1950	6,665,863	214,489	257,292	522,515	2,396	<100,000
1960	6,703,108	1,816,779	1,175,301	1,352,728	165,094	
1970	6,550,150	3,375,822	2,225,699	1,650,000	3,130,000	
1979	8,485,826	3,930,153	3,162,276	1,029,036	6,288,330	
1980	6,436,068	3,460,092	2,921,650	937,048	7,176,250	
1990	6,069,431	4,239,356	2,662,502	1,260,768	10,213,113	964,607
1998	6,388,101	5,132,746	2,680,270	1,602,026	8,010,843	1,282,726

Source: *Automobil-Revue und Catalog*, various years. Due to differing accounting methods, figures not all directly comparable.

Figure 157. Buick Riviera hardtop coupe, 1970. Compared to the excesses of the 1950s, by 1970 the average American car, as represented by Buick, showed more restraint when it came to chrome trim. However, with overall lengths around 570 cm (19 ft) and weights of 2000 kg (4400 lb), they still had a majestic presence. In 1970 (i.e., before the first energy crisis), Buick introduced a 7.5-liter (455-cu.-in.) engine, the largest and thirstiest V8 in corporate history. V8 engine, 7456 cc, 370 hp (approximately 272 kW).

However, steadily growing numbers of motor vehicles, especially in urban centers, precluded any consideration of the automobile as an isolated object, divorced from its environment. Mass motorization brought with it negative side effects. The auto industry itself ignored these, thereby forcing the American government to take regulatory action.

Positive crankcase ventilation, introduced in the mid-1960s, was followed by California's specific passenger car exhaust emissions limits as of 1968. Initially, car makers applied add-on solutions such as thermal reactors, catalytic converters, and exhaust gas recirculation. These methods consume oxygen in operation, and additional raw materials and energy in their manufacture. Potentially, they increase fuel consumption and reduce power, although in practice surprisingly good results could be achieved.

Exhaust gas post-treatment was followed by external engine changes—for example, installation of fuel injection systems to replace carburetors. As a matter of convenience, manufacturers went from carburetors directly to electronic fuel injection systems, without detouring through the older technology of mechanical injection and thereby avoiding the need for engine-driven metering systems. Volkswagen took this approach with the Type 3 (Volkswagen 1600), because its four-cylinder boxer engine could not meet U.S. emissions standards with two side-draft carburetors, and conversion to a mechanical injection pump would have required major design changes.

The U.S. version of the Volkswagen 1600, 1967/68 (Figure 158) was the first mass-produced midrange car to be fitted with electronic fuel injection—under duress, for without pressure

*Figure 158. Volkswagen 1600 (Type 3), 1967. Beginning in 1967, Volkswagen intro-
duced the world's first mass-produced, electronically fuel injected cars—the Type 3
(1600) and Type 4 (411E, 412E). Initially, only 1600 models intended for export to
the United States (shown here) received the Bosch D-Jetronic injection system. As of
June 1968, it was made an option for the German market as well. Fuel economy and
performance remained virtually unchanged from the carbureted version. Four-
cylinder horizontally opposed engine, 1584 cc, 54 hp (40 kW).*

from U.S. lawmakers, this and other improvements probably would not have been introduced
until much later. For the first time in the history of the automobile, government influenced
layout and design of passenger cars for reasons unrelated to taxation. To be able to export
their wares to the United States, foreign car makers had to meet American emissions standards, which unwillingly prompted similar standards in their home countries.

Simultaneously, fuel injection technology achieved a new attribute: it was no longer employed
primarily to increase power output, as had been the case in its development years, but rather to
reduce exhaust emissions. From the late 1960s, sanitizing the exhaust and reducing fuel
consumption unfolded as primary goals for the future. Ecological advantages and expected
advances and price reductions in semiconductor technology, as well as its perception as a
generally more elegant solution, suggested that Volkswagen would not be the only auto maker
to build electronically fuel injected engines.

And so it was. Beginning in 1968, the Volkswagen Type 3 was joined by the Mercedes 250 CE
and 280 SE 3.5, the Opel Admiral/Diplomat, Volkswagen-Porsche 914, and Citroën DS 21. By
the spring of 1976, six German, three British, two French, and both Swedish car makers
employed Bosch electronic fuel injection.

The Bosch D-Jetronic system of the Volkswagen 1600 was in turn based on licenses obtained
from an American firm, Bendix Corporation, probably the first to explore the possibilities of
electronic controls for injection systems. In 1953, Bendix presented a Buick V8 fitted with the
firm's "Electrojector" system to the auto industry. Chrysler was impressed enough to offer

several 1958 models with electronic fuel injection. For approximately $640 in added cost, intrepid customers received about 10 extra horsepower "...and no end of trouble" (*Special Interest Autos*, Vol. 112, 1989). Chrysler had to recall the few injected cars in circulation and refit them with carburetors. However, Bendix continued to develop the system, integrating advances made in other areas of electronics, and by 1965 had achieved such a high level of technical refinement that Bosch, facing time pressure with its 1965 contract to deliver electronic injection pumps to Volkswagen, found it more expedient to take out a Bendix license.

The second regulatory measure undertaken by government agencies developed into a massive blunder perpetrated by several unqualified bureaucrats. Facing an annual rise in highway traffic fatalities (49,000 in 1965), the Department of Transportation (DOT) contracted with General Motors and two firms not related to the auto industry for the development and construction of so-called Experimental Safety Vehicles (ESVs).

Completed in 1971–1972, these ESVs (Figure 159) were bulbous and even heavier than American production cars but, with their huge front and rear overhangs, promised survivability in frontal crashes up to 80 km/h (50 mph). It also happened that in the decade of the 1960s, American car production dwindled from approximately 52 to 30 percent of worldwide production, while imports to the United States soared. Foreign manufacturers of small cars feared a new form of exclusionary competition. To retain their share of the U.S. market, European and Japanese firms of necessity joined the "safety festival."

Figure 159. AMF ESV, 1971–1972. The Department of Transportation (DOT) issued development contracts for an Experimental Safety Vehicle (ESV) to AMF Inc., Fairchild Industries, and General Motors. The resulting "safety cars" from AMF (shown here) and Fairchild had automatically extending front bumpers, airbags instead of seatbelts, extensive interior padding, and periscopes instead of rear-view mirrors. The chassis layout remained conventional. V8 engine (Chevrolet), 5733 cc, 157 hp (116 kW).

Safety car mania subsided immediately after the 1973–1974 energy crisis, but not until millions, perhaps billions, had been invested senselessly. After all, the basic proposition of the concept was seriously flawed. The ESV program concerned itself with passive safety and only passive safety. In the estimation of renowned safety engineer Béla Barényi, passive safety comes in last place behind active safety and perceptive safety (e.g., emergency flashers, reflective license plates). Since the 1930s, Europeans had led the way in active safety—in other words, road manners and handling. Since 1959, they also led in passive safety—the reduction of the effects of an accident. In that year, Mercedes introduced its 220 (Figure 160), the first car to have a rigid passenger cell and decreasing stiffness toward the extremities of the car. Today, all cars, including American products, are built to Barényi's "crush patent."

Figure 160. Mercedes-Benz 220, 1959. As outlined in Barényi's "crush patent" of January 21, 1951 (German patent number 854,157), in turn based on two of his earlier patents dated 1945 and 1946, front and rear sections should deform in the event of an accident; however, the passenger cell should retain its shape. Doors, removed here to permit observation of the crash dummies, should remain operable by hand even after an accident. Six-cylinder engine, 2195 cc, 95 hp (70 kW).

The key phrase has already been mentioned: energy crisis. This was a result of the Yom Kippur War between Egypt and Syria versus Israel in October 1973. For the first time, the Arab states used oil as a political weapon. They raised oil prices, throttled the supply, and instituted an embargo against the United States and the Netherlands as punishment for their support of Israel. The Organization of Petroleum Exporting Countries (OPEC) had been founded in 1960 as a response and defense against multinational oil companies. Its pricing policy mirrored the Arab embargo, turning the organization into a producer cartel of overpowering strength, and perfectly willing to use that strength. Crude oil prices, which in 1970 had been around $1.40 per barrel, quadrupled by 1973. The Western oil companies took advantage of the shortage and raked in massive profits.

The German Federal Republic reacted to the oil crisis by reducing energy consumption, four "carless" Sundays in November and December 1973, and 100-km/h (62-mph) speed limits on the Autobahn. With one stroke, the public was made painfully aware of its dependence on crude oil imports and the limited nature of oil reserves.

Rising oil prices that drew money out of industrialized nations, a weakening U.S. dollar, the collapse of the worldwide currency system in 1973, and increasing competition from goods made in Japan or low-wage Third World countries all led to the most serious economic crisis in the West since the end of World War II. As the engine that powered the entire world economy, the United States also suffered from huge deficit spending, the Vietnam conflict, the Watergate affair, and racial strife. The worst recession since the world economic crisis of 1929 manifested itself in high unemployment (7.9 percent in 1976) and high inflation (12.2 percent in 1974).

In 1974 and the years that followed, OPEC continued its policy of oil price increases. The Iranian revolution of 1978–1979 unleashed a second oil crisis, which raised crude oil prices to almost $23 per barrel. In 1981, crude prices reached their highest level to date—$34 per barrel.

In response to the first oil crisis, the U.S. government imposed fuel economy requirements on top of emissions and safety standards. The prescribed corporate average fuel economy (CAFE) (i.e., the average fuel economy of all vehicles made and sold domestically by a given manufacturer), had to exceed 20 mpg (below 11.75 liters/100 km) by 1980, and 27.5 mpg (8.55 l/100 km) by 1985.

Unintentionally, this requirement had an enormous effect on automotive technology, the American economy, and indeed the entire American way of life. No manufacturer was able to meet CAFE with its 1975 models. In round numbers, even the subcompact Chevrolet Vega, a small car by U.S. standards, got no better than 17 mpg (14 l/100 km). A midsize car such as the Oldsmobile Cutlass consumed a gallon for every 14 miles (17 l/100 km), while the Cadillac Fleetwood happily gulped down a gallon every 9 miles (26 l/100 km).

For years, "with suicidal shortsightedness" (*Süddeutsche Zeitung*, Feb. 21, 1980), the American auto industry had ignored the energy crisis. Instead of developing smaller cars with moderate fuel economy, Detroit continued to build land yachts with voracious fuel appetites. Worse, the Massachusetts Institute of Technology (MIT) characterized American industry in general as lagging behind Japan and other industrialized nations in terms of productivity, product quality, and innovation. It was a well-known fact that the build quality of American cars was not up to the standards set by European or Japanese cars.

Car makers reacted, belatedly, with three measures: a crash diet for their massive cars, installation of more economical diesel engines, and new designs for smaller and more economical cars.

In 1981, the United Auto Workers forced through a "self limiting agreement" aimed at curtailing Japanese imports. Beginning in 1985, American and Japanese auto makers began joint-venture production—in effect, an admission by American firms that they could not meet the demand for smaller cars without help from their Japanese competition.

Weight reductions were relatively easy to accomplish. For example, Cadillac shortened its 1977 Fleetwood by almost 13 in. (32 cm) and downsized its V8 engine from 8.1 to 7 liters, which pared 430 kg (948 lb) of curb weight. Oldsmobile introduced the first American production passenger-car diesel in 1977, followed in 1978 by Cadillac and other General Motors models. A new vehicle generation appeared at the end of the decade—the 1979 General Motors X cars (Figure 161), the 1980 Chrysler K cars, and in 1981 Ford's "world cars," the Escort and the Lynx.

All of these cars exhibited modern technology (front wheel drive, transverse engines), decent exterior dimensions, low weight, good fuel economy, higher quality, and better space efficiency, comparable to their European and Japanese rivals. By 1983, they had returned their companies to profitability, after billion-dollar deficits for all four American concerns in the previous years. Chrysler survived only because of government-secured loans, withdrawal from Europe, and radical restructuring under chairman Lee Iacocca.

Figure 161. Chevrolet Citation, 1980. In 1975, General Motors began development of its "X car" platform—family cars in a European format, with front-wheel drive and transverse four- or six-cylinder engines. Weight (approximately 1100 kg [2425 lb]), fuel economy, and drag coefficient (c_d around 0.42) were comparable to European cars. Four-cylinder or V6 engine, 2474 or 2855 cc, 91 hp (67 kW) or 115 hp (85 kW).

One firm, however, did not survive: American Motors Corporation/Jeep, formed in 1957 by the merger of Nash and Hudson. Although AMC was the first to turn its back on "gas-guzzling dinosaurs" (a phrase often mistakenly attributed to AMC chairman George Romney) and offer compact cars, Renault acquired a controlling interested in AMC stock in 1980. In 1987, control of AMC passed to a resurgent Chrysler Corporation. Particularly painful for American industry was the fact that in 1980, Japan took first place in production numbers and, except for two years, held that position until 2000. The United States had lost ground, both quantitatively and qualitatively.

The oil crises also had a negative effect on production and registrations in Europe, although not to the same extent as in North America. Along with American requirements and increased environmental awareness, European auto makers were forced to build cars appropriate for these changed conditions. After years of reconstruction and a period of profit maximization, in which engineering progress (at least in Germany) was doled out in minuscule portions, technical problems again held top priority.

The 1960s saw its share of new ideas, even if those ideas were not immediately embraced by the German industry. For example, Fritz Ostwald proposed negative steering offset (known as KLR, for "Kursstabilisierender Lenkrollradius"—course-stabilizing steering offset) in 1958, to achieve better stability even on split-coefficient surfaces (e.g., ice versus concrete or asphalt) or in the event of one-sided brake failure. The idea met with little interest in Germany; however, General Motors made good use of it on the 1965 Oldsmobile Toronado (Figure 164), the first car to employ a negative steering offset. Audi did not follow suit until 1972. Since then, negative steering offset has become standard practice throughout the automotive industry.

Long gestation periods also were encountered by some of the safety features suggested by Béla Barényi, such as knee bolsters (1964), safety steering (1960), and lenticular roof (1956), introduced by Mercedes in a modified form on the hardtop for the 1963 230 SL (the "pagoda roof") convertible. Otherwise, Mercedes held fast to the single-pivot swing axle, DKW remained faithful to the two-stroke, and Volkswagen kept its rear-engine layout. German auto makers introduced disc brakes years after their British and French competition. The rubber timing belt pioneered by Glas in 1962 and air suspension remained glimmers of light in an otherwise lackluster era.

Passenger car air suspension, introduced by Borgward and Mercedes, proved to be more reliable than the systems offered by General Motors, Ford, Chrysler, and Rambler as of 1957. These had been rushed to production under tremendous time pressure, proved to be not fully developed, and hence disappeared from the scene by 1960. In the same year, Borgward offered the first European car with air suspension (Figure 162). Mercedes followed a year later with the 300 SE and retained air suspension for the "Grosser Mercedes" 600 until 1981. In 1984, Lincoln overcame the inability of air suspension systems to absorb short, hard shocks by installing electronic controls. For years, the Lincoln Mark VII was the only production car to use air suspension, until Mercedes returned in 1998 with an admittedly much more complex system in the new S class. The future probably will see air suspension in off-road leisure vehicles as well as in passenger cars.

Figure 162. Borgward P 100, 1962. Given the catchy "new German" name "airswing," the P 100 employed air suspension jointly developed by Borgward, Firestone-Phoenix (rolling-lobe air springs), and Bosch GmbH (compressor, air reservoir, and control valves). The inability of Borgward's (and also Mercedes') air springs to absorb short, sharp jolts was criticized even then. The Borgward also was available with conventional steel springs. Six-cylinder engine, 2238 cc, 100 hp (74 kW).

Less satisfactory was body styling. Depending on the attitude of the manufacturer, this tended either toward the "German rustic furniture" school or recalled the Baroque excesses of Detroit. One notable exception was the Porsche 911, although it should be noted that it is easier to design an attractive, convincing two-seat sports car than a four- or five-seat sedan. NSU was able to achieve this difficult feat in 1967 with the Ro 80 (Figure 163). With its sloping nose, rising beltline, tall trunk (wedge shape), three side windows, and good aerodynamics, the Ro 80 influenced worldwide passenger car body design through today. Claus Luthe's design of the Ro 80 wedge finally displaced the Italian design ethic that had held sway in the 1950s and 1960s. Quite possibly, the wedge will in turn be replaced by French styling *à la* Renault, typified by "Avantime" and other designs by Patrick LeQuement.

The novel features of the Ro 80 did not end with its body shape. Powered by a Wankel rotary engine (see Appendix), the Ro 80 was the world's first Wankel-powered sedan (see chapter on alternative fuels, engines, and drive systems). Moreover, front-wheel drive, long wheelbase, and front and rear disc brakes, plus well-defined crush zones, a rigid passenger cell, and a

Figure 163. NSU Ro 80, 1967–1977. Today, approximately 35 years after its introduction, the body shape designed by Claus Luthe appears fresh and modern. With a drag coefficient (c_d) of 0.355, the Ro 80 was for years one of the slipperiest shapes on the road, to say nothing of its aesthetics. In all, 38,052 examples were built. Two-rotor Wankel engine, chamber volume 2×497.5 cc, 115 hp (85 kW).

steering box set high and well back, as suggested by Barényi, provided above-average levels of active and passive safety.

In its day, the Ro 80 was the largest German front-drive car. Because of the success of the Mini, front-wheel drive advanced along a broad front throughout the European auto industry. Between 1961 and 1969, Fiat, Renault, Lancia, Ford of Germany, Triumph, Peugeot, and Simca switched from rear- to front-wheel drive. To everyone's surprise, the aforementioned Oldsmobile Toronado joined the club in 1965 (Figure 164), only the fourth American front-drive production car after those of Erret Lobban Cord (between 1929 and 1935) and the 1929 Ruxton. With its 390 SAE horsepower, the Toronado proved that front-wheel drive was not limited to small and midsize cars. It was joined a year later (1966) by the technically similar Cadillac Eldorado and much later (1979) by the Buick Riviera.

After acquiring Auto Union from Daimler-Benz in early 1965, even Volkswagen, fixated as it was on the rear-engined Beetle, had two front-drivers in its lineup—the DKW F 102 and the 72-hp Audi. By the time Volkswagen absorbed NSU in 1969, it had a colorful palette of models at its disposal: front-drive cars (Audi, NSU), all-wheel drive (DKW Munga), and rear engine/rear drive (Volkswagen, NSU); air-cooled (Volkswagen, NSU) and water-cooled (Audi, NSU) powerplants; engines with two (NSU), three (DKW), and four (Volkswagen, Audi, NSU) cylinders; two strokes (DKW) or four strokes (all others); and, finally, the Wankel rotary engine in the NSU Ro 80.

Figure 164. Oldsmobile Toronado, 1966. With the Toronado, the General Motors Oldsmobile Division sought to combine a powerful engine with exceptional handling and stability. With the first application of negative steering offset (see Appendix), the Toronado had front-wheel drive and an unusual drivetrain: the torque converter, bolted to the back of the engine, was connected to the automatic transmission and planetary gear differential (both located alongside the engine) via chain drive. V8 engine, 6965 cc, 385 hp (approximately 283 kW).

Such a broad spectrum of model variations rules out any sort of rational manufacturing strategy. Because Beetle production costs were also too high, Volkswagen was awash in a sea of red ink, despite rising sales numbers. As luck would have it, such was the situation in the landmark year 1972 as Volkswagen was celebrating the completion of chassis number 15,007,034 on February 17, the car that put Volkswagen ahead of the Ford Model T as the most-built car in history. (As it turned out, the Volkswagen celebrations were premature because Wolfsburg had not taken into account the 400,000 "Tin Lizzies" built in Canada).

Neither the Beetle nor the other Volkswagen models had what it took to ensure the survival of the largest car maker in Europe. Prerequisites such as low manufacturing and maintenance costs, low weight, good performance, high comfort, attractive shapes, practicality, and high active and passive safety demanded an all-new design to achieve worldwide acceptance, high production numbers, and a long production life. The result was the Audi 80, which commenced deliveries in September 1972, and its sister model, the Volkswagen Passat (July 1973).

With front-wheel drive, course-stabilizing steering offset, diagonally split braking circuits, semi-beam rear axle, body sheet metal thickness determined by finite element methods (FEM), floating-caliper disc brakes, and overhead cam engine, the Audi 80 and Passat formed the basis for future Audi and Volkswagen models. This includes the Golf (Figure 165), which one day would achieve a reputation to rival that of the Beetle.

Figure 165. Volkswagen Golf, 1974 to the present. Although built in northern Germany, in terms of styling the Golf has an Italian father (Giorgio Giugiaro) and, with its hatchback layout, Italo-British ancestors in the 1959 Mini, 1959 Austin A 40 Countryman, and 1964 Autobianchi Primula. The Golf (originally sold as the Rabbit in the United States) has surpassed the Beetle as Wolfsburg's most important product, setting the standard for an entire class of compact cars. Four-cylinder gasoline and diesel engines, 1093 to 1588 cc, 37 to 81 hp (27 to 60 kW) (1979).

In case German customers suddenly developed a craving for small Italian and French products, Audi developed a modern compact car of its own. The OPEC oil price hikes of 1973–1974 resulted in accelerated introduction of the Audi 50 (Figure 166), which went into production in September 1974.

Although small cars had never played as significant a role as in neighboring countries, with the Audi 50 Germany could respond to the oil crisis with a product superior to its market competitors, the Peugeot 104 C, Autobianchi A 112, and Mini. It offered not only a rear hatch and four viable seats, but in the event of an accident, had collapsible longitudinal members ahead of the firewall; a rigid perimeter around the floorpan, which diverted side impact forces into deforming the central tunnel; a rigid structure consisting of hatch frame, C pillar, and seat transverse member; and a plastic instrument panel with deliberate break points outboard the likely head contact zones. The Audi 50, also marketed as the Volkswagen Polo from 1975 onward, is an excellent example of how the design of a full-featured, safe compact car with market-driven pricing represents a greater technical and logistical challenge than a luxury car overloaded with technology for the sake of technology.

Rising oil prices gave new appeal to diesel engines for passenger cars because their thermodynamic efficiency is higher and fuel consumption lower than those of a conventional gasoline engine. Before the oil crisis, Mercedes, Opel, Peugeot, Austin, Land Rover, and Datsun

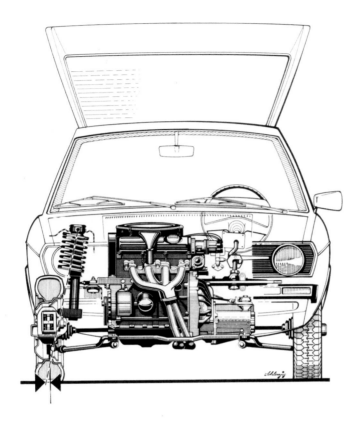

Figure 166. Audi 50, 1974. With an overall length of only 3.49 meters (11.45 ft), the Audi 50 was one of the most compact small cars in a class populated by larger, older competitors—and the more modern 1973 Peugeot 104 C and 1969 Autobianchi A 112. Because of its rear hatch and four usable seats, the Audi was superior to these. This layout was made possible by the first use of a transversely mounted engine, designed by Martin Probst, in an Audi product. Note negative steering offset (arrows). Four-cylinder engine, 1093 cc, 50 or 60 hp (37 or 44 kW).

offered passenger cars with diesels. Potential customers were lukewarm toward diesels because of the diesels' leisurely acceleration, noticeable lack of power on grades, annoying noise levels, and image problems, at least in Germany. Lacking motivation and competition, no significant progress was made in the diesel field. Worldwide, diesels accounted for only 0.6 percent of production in 1970.

Despite its lack of prior in-house experience, Volkswagenwerk explored the potential of diesels. The decisive factors were the energy crisis and fuel economy and emissions standards in the United States, Volkswagen's most important export market. In short order, under project leader Peter Hofbauer, the Passat gasoline engine was converted to diesel operation and offered in the

Golf (Rabbit) as an alternative powerplant. The prechamber diesel made an immediate impression on automotive journalists and the public alike. With a specific output of 16.1 kg/hp (for comparison, the Mercedes 200 D had 25 kg/hp), the Golf Diesel could be driven almost as briskly as the gasoline model. In terms of fuel economy, maximum engine speed, acceleration, top speed, price, and power per liter, the Golf Diesel outdistanced all of its competitors, thereby giving a powerful stimulus to further development of passenger car diesels.

In retrospect, the first oil crisis of 1973–1974 acted as a catalyst for developments that were already underway: reduction of fuel consumption and exhaust emissions, as well as increased active and passive safety. The oil shortage led to the widespread realization that vehicle traffic had to be viewed in a global context. In the aftermath of the oil crisis, "The Limits to Growth," a report published by the Club of Rome in 1972 and at first widely ignored, prompted individual citizens and organizations to question unrestrained expansion of industry and consumption, and unbridled use of land and resources. In Germany, this led to ballot initiatives and the rise of the Green Party. Begun in the 1970s, government-mandated environmental protection by the end of the decade had shifted its focus from the mere removal or minimization of existing environmental damage to prevention of future environmental problems. In the 1980s, even the "jobs versus environment" controversy yielded to the recognition that environmental protection definitely creates jobs as well.

The Japanese Challenge

From the 1960s, Europe and North America have been confronted by industrial products from countries which, although far removed from Western thought and manufacturing traditions, have adopted Western rationalism into their own production methods.

Within only 15 years, Japanese car makers put an end to approximately 80 years of status quo within the auto industry and challenged the position of American and Western European car makers as cornerstones of the economy. In this regard, the auto industry is quite different from other economic sectors that had already been paralyzed by Japanese competition: watches, cameras, pocket calculators, motorcycles, and copying machines.

Employing manufacturing methods and export strategies that seemed to guarantee success, Japanese auto makers first conquered markets in countries lacking domestic auto industries of their own and, shortly thereafter, the car-producing countries. Although Japanese motor vehicles (passenger cars, trucks, and buses) could claim only 0.9 percent of the world market in 1956, by the 1980s and 1990s, Japanese passenger cars alone settled on a market penetration of 20 to 25 percent, with a spike of 29.4 percent in 1991 representing 10,088,430 cars and station wagons. For comparison, Germany produced 4,277,504 units (12.5 percent) and the United States 5,439,879 (15.9 percent of the market).

Japan's success in such a short time is all the more surprising if we consider its origins. The first Japanese gasoline-powered car was built in 1907. By 1920, Japan had approximately

9,000 cars, including imports. In the aftermath of the great 1923 earthquake, which destroyed the railway and streetcar networks, Ford and General Motors received permission to build assembly plants in 1924. Chrysler followed in 1927. By 1935, Japan had 35,000 vehicles, which prompted the government to ban imports of components and thereby protect the domestic auto industry from foreign influence. The American assembly plants simply withered away.

However, Japanese car makers were unable to deal with the resulting dilemma. Although they had sovereignty in their home market, they also lacked sufficient engineering expertise to go it alone. After the nation and its markets were reopened by American forces in 1945, the Japanese auto industry again (or better said, still) found itself in the position of licensee. Whereas in the 1930s Toyota had copied the Chrysler Airflow (Figure 106) and Datsun/Nissan did the same with the American Graham six-cylinder and the British Austin Seven (Figure 78), now Datsun took a license from Austin (A40/A50), Isuzu from Rootes (Hillman Minx), Hino from Renault (4 CV), and Mitsubishi from Willys-Overland (Jeep).

The first attempts by Japanese industry to free itself of foreign licenses occurred with its third generation of cars, which characterized the 1950s. Regulatory requirements resulted mainly in 3.20-meter (10.5-ft) long bonsaimobiles with a displacement limit of 360 cc, usually rear-engined with simple suspensions. These were unsuitable for export because they were too small, too primitive, and too ugly.

Around 1961, the Japanese ended their era of imitation, rebuilt or expanded their production facilities, and forced passenger cars of their own design on a domestic market at the same time that its road network was being developed. Several representatives of this fourth generation, such as the Toyopet Corona, Hino Contessa, and Datsun Bluebird, were characterized by definite "Italian" styling features. In the interest of exports, in following years this styling would have an ever-stronger American influence. Around 1980, the Japanese again swung to European fashion, in this case German body styling.

Although the first export attempts made by Datsun and Toyota to the United States failed (1957 and 1959, respectively), the actual Japanese export offensive began in 1965, with improved fourth-generation models. Production numbers rose from 268,784 units in 1962, as the new models began to take hold of the market, to 2.058 million in 1968. This put Japan ahead of England, France, and Italy, and behind only the United States and Germany in the global pecking order. In 1971, Japan passed Germany for second place in the production standings. In 1974, it passed Germany in exports, and in 1980 it pulled ahead of the United States. In only 20 years, Japan had gone from a regional producer to a global automotive superpower.

The meteoric rise of Japan was based on an entire series of ethnic, social, and economic/political peculiarities, but also on an inability or unwillingness by Western firms to rationalize their production and tailor their products to customer needs.

The Japanese, on the other hand, showed complete mastery of both disciplines. On the one hand, they increased the number of mechanized and automated operations, introduced quality assurance early in the manufacturing process, instituted work groups with "flat" hierarchies, increased outsourcing while reducing the depth of their own manufacturing activities ("lean production"), and cut warehousing costs with "just in time" delivery of components and assemblies.

The primary goal was and remains improved productivity by means of cost reduction. In numbers, productivity of Nissan's British plant, which commenced operations in 1986, was 40 percent better than the average of European car plants. It took less than 17 hours to produce a car in Japan, approximately 25 hours in the United States, and slightly more than 36 hours in Europe (1992).

The Japanese also created a demand for niche products which, often as not, grew to become major profit centers. Examples include all-wheel drive for small pickups (minitrucks), introduced by Subaru in 1978 (Figure 167) and later found on station wagons and sedans; the Mitsubishi Super Space Wagon (Figure 168) of 1979, which unleashed the minivan craze in Europe and the United States; and affordable roadsters, minivans, three-cylinder four-stroke engines, five-speed transmissions, four-wheel steering, multi-valve technology, and navigation systems.

To some governments, the deluge of cars from the Far East seemed so ominous that they imposed import restraints. Beginning in 1955, Italy limited imports to 2,200 units, then 3,300 in 1988, with an additional 5,000 from third nations. Beginning in 1978, France limited the Japanese

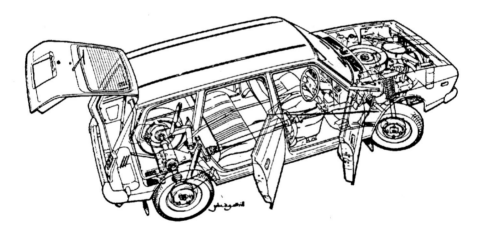

Figure 167. Subaru 4WD, 1979. With the exception of the British Jensen FF, a luxurious, high-performance coupe first shown in 1965, four-wheel drive was reserved for off-road or special-purpose vehicles and tractors. Today, customers expect even ordinary passenger cars to offer good traction under slippery road conditions, balanced handling, and short braking distances. The Subaru had a front-mounted horizontally opposed engine similar to that of the 1957 Goliath, with full-time front-wheel drive and selectable rear-wheel drive. Four-cylinder engine, 1595 cc, 67 hp (49 kW).

Figure 168. Mitsubishi Super Space Wagon, 1979. Although they might lack prominent engine hoods, notchbacks, and low overall height, roomy and attractive (mini-) vans can provide good aerodynamics without resorting to delivery van design ethics. Predecessors of the Mitsubishi van, which established a new class of passenger car, were the Stout Scarab (United States, 1935), Fiat Multipla (Italy, 1956–1960), and autonova fam (Germany, 1965, Figure 220). Four-cylinder engine, 1783 cc, 90 hp (66 kW) (1983).

to 3 percent of its market. Great Britain followed in 1980 with 11 percent. As of 1981, the United States set limits on Japanese imports of 1.68 to 2.3 million units. Of the major car-producing nations, only Germany kept its market open; Japanese market penetration measured 6 percent in 1979, 15 percent in 1990, and 11 percent in 1999. The first Japanese import to land on Hamburg docks was a Honda S 800 (Figure 169) in 1967.

The restrictive efforts of all these nations proved fruitless; if anything, quite the opposite resulted. In 1980, in all four countries, domestic auto production dropped even more than in Germany's more liberal market. A study by Daimler-Benz (*Automobil Revue*, Bern, Switzerland, September 22, 1988) concluded that import restrictions did more harm than good. The Japanese saw themselves as forced into basic restructuring. To circumvent restrictions, they took two different approaches. First, they began to produce cars locally. By 1987, Honda, Nissan, Toyota, and Mazda had established plants of their own in the United States. Other firms entered into joint ventures with American car makers. Second, they shifted their model lines toward more expensive products to maximize profits in the face of restrictions that set limits on numbers rather than value. To more quickly free themselves of their "cheap car" image, they introduced new marques: Lexus at Toyota, Acura at Honda, and Infiniti at Nissan.

When Japan shifted its product lines upscale, it left a vacuum at the lower end of the range, which soon was filled by Korean and South American makers, but also the tiny Yugoslavian-built Yugo. North American makers themselves were either unwilling or unable to enter that niche.

Figure 169. Honda S 800, 1967. Similar to its predecessor, the S 600, the Honda S 800 was powered by a small but potent engine. Double overhead cams and four carburetors squeezed 67 hp and 7500 rpm out of the 800-cc engine. At 84.7 hp/liter, its specific output was higher than that of the contemporary Porsche 911 S (80.4 hp/liter) and all of the British sports cars that were its intended competition. Four-cylinder engine, 791 cc, 67 hp (49 kW).

In Europe, German car makers gained the most from the protectionist policies of France, Britain, and Italy because in those countries, the Japanese had to be content with a much smaller presence. On the other hand, despite protectionism, France and Britain saw only minimal gains in their own home markets and even suffered an overall drop in sales because the Japanese concentrated on those Western European markets that remained open to them.

Despite these facts, Britain, Italy, France, Spain, and Portugal insisted on import restrictions even after the creation of Europe's single market on January 1, 1993. The Japanese accepted an import quota of 1.23 million units and shipments from transplants of 1.2 million units through 1999, although under the General Agreement on Tariffs and Trade (GATT), they were not required to accept self-imposed restraints without compensation.

After all, Europe had followed the situation in the United States with some concern where, by 1991, Japanese imports and transplants accounted for approximately 30 percent of the market and had severely weakened the American auto industry. Now, they feared an assault on the European market. After Japan had conquered its Asian neighbor markets in the 1970s and the United States in the 1980s, it appeared that the target for the 1990s might well be Europe.

As in North America, Japan reacted by erecting plants in Europe. By 1997, it had established eleven plants within the European Union and one in Hungary, plus nine research and development centers, some in cooperation with European partners. Import restrictions of 1.15 million units or less, effective from 1995, no longer worried the Japanese because they could circumvent these with vehicles produced by their transplant factories.

Meanwhile, the economic and political situation had shifted again, and not only because of the worldwide recession of 1992–1993. Not counting occasional minor crises, the early 1990s saw the first decrease in production, export, and registrations in Japanese automotive history. The Far Eastern auto industry, once a seemingly unstoppable juggernaut, suffered setbacks from which, at this writing, it has yet to recover.

Japan lost some significant advantages over foreign competitors, among these the undervalued Yen. Wages, formerly low, reached American levels by 1992. Low-interest loans, which encouraged investment in the years of stock market speculation (1986–1989), suddenly dried up. Because of traffic congestion in metropolitan Tokyo and the Kawasaki/Yokohama industrial zone, just-in-time deliveries could not be maintained. Too many production locations and excess capacity forced plant closings and layoffs. The system of lifelong employment became a thing of the past. Nissan and Mitsubishi reported net losses.

The Southeast Asian financial crisis of 1997–1999 had a dramatic effect, as did the resurgent European auto and supplier industry, which had made good its disadvantages in cost and production technology and, to some extent, even outpaced its Japanese rivals. This new self-confidence was expressed in capital investment and alliances with once-impregnable Fortress Japan. Volvo Trucks bought into Mitsubishi Motors, Bosch increased its participation in strategically important supplier Zexel Corporation, Goodyear Tire took control of SP/Dunlop, and General Motors allied with Fuji Heavy Industries (Subaru). In 1996, Ford increased its holdings of Mazda from 25 to 33.4 percent. In 2000, DaimlerChrysler acquired 34 percent of Mitsubishi Motors (passenger cars and light commercial vehicles). It is only a matter of time before Nissan Diesel becomes a takeover target. To date, the most spectacular acquisition is Renault's 36.8 percent stake in Nissan Motors. Since 1999, the French, who only a few years earlier had been on the financial brink themselves and a prime takeover candidate, have been calling the shots at the second-largest car maker in Japan.

Similar to Rome, Spain, and England before it, Japan has lost its myth of invincible economic power. In other words, this is a completely natural process. However, Japan will surely recover from this shock and emerge even stronger. European-Japanese and American-Japanese cooperation will continue to make the automotive world a little smaller.

The Automobile as a Consumer Good

1980 to 2000

New Horizons through Electronics

A rising standard of living, expansion of the road network, land development, centralized schools in rural areas, and division of extended families into small households led to a rapid increase in individual transportation. In 1953, 52 percent of West German citizens shared cars with others; in 1958, the percentage dropped to 21, in 1988 to 2.1, and in 1999 to only 1.9 percent. This represented American conditions—30 years late, and in a nation (West Germany) with only 1/37th the land area.

The rising number of vehicles demands greater concentration on traffic conditions (i.e., the car itself enjoys less attention). It is reduced to a consumer good, always ready, and the only requirement is that it function, a situation that even in the 1960s was by no means taken for granted. On the other hand, mass motorization demanded ever more environmentally responsible automobiles to preserve an already threatened standard of living.

As always, American firms were at the forefront of providing appropriate and more comfortable transportation. The American auto industry introduced the electric starter that eliminated hand cranking. It also equipped its cars with power steering, synchromesh, and automatic transmissions because it saw no reason why a driver needed to exert himself or herself while steering, shifting, or engaging the clutch. However, in the course of continued development, particularly in the application of external requirements such as emissions, it was soon realized that mechanical, hydraulic, and even electrical systems had reached their limits, and that future regulations could not be met without additional assistance. Engineers, for ages associated with and trained in the field of mechanical engineering, had no choice but to adopt the electronic solutions offered by suppliers, which had already made inroads in other economic sectors. In layout and material condition, the automobile, once an easily comprehended product of the mechanical engineer's art, developed into a cutting-edge technology carrier, with contributions from the electronics and aerospace industries.

To meet ever-increasing electrical power demands even at idle, in the early 1960s AC generators (alternators) with semiconductor diodes appeared in commercial and rail vehicles, soon followed by American and European cars in the higher price classes.

The same time frame saw the inevitable shift from conventional coil ignition to electronic breaker-controlled, later strictly breakerless transistorized coil ignition (TCI) as well as thyristor or high-voltage capacitor discharge ignition (CDI; see Appendix, Ignition systems). Electronic ignition systems provide a more powerful spark, with resulting improvements in exhaust emissions, fuel economy, power output, cold starting, and warm-up behavior. In addition, higher engine speeds are possible.

The first production car with electronic ignition may have been the NSU Spider. It went into production in September 1964. With the first application of a Wankel engine to a production car, it also employed breaker-controlled high-voltage capacitor discharge ignition (CDI; Figure 170).

Figure 170. NSU Wankel Spider, 1964. The Spider was powered by a single-rotor Wankel engine, mounted at the rear in unit with transmission, differential, and single-disc dry plate clutch. Of interest is the capacitor discharge ignition (CDI) system; its cam, ignition points, and dual capacitors are visible on the extended nose of the eccentric shaft. The actual CDI unit and its circuit board can be mounted anywhere in the vehicle. Single-rotor Wankel engine, 500-cc chamber volume, 50 hp (37 kW).

While capacitor discharge ignition remained limited to rotary engines and high-speed reciprocating piston engines for sports car applications, transistorized coil ignition (TCI) was well suited for ordinary production piston engines. The first breakerless ignition system fitted to a German car was found in the Glas 3000 V8, which appeared under the BMW banner in September 1967.

After formulating emissions and fuel economy regulations, in the 1970s American regulations mandated ignition systems that would operate without maintenance for 50,000 miles (80.465 km). This reduced-maintenance argument forced the introduction of breakerless electronic ignition systems. Around 1978, BMW, Chrysler, Citroën, Fiat, Lancia, Leyland, Mercedes, Peugeot, Porsche, and Volvo employed breakerless electronic ignition, universally of the TCI variety.

Subsequent development of transistorized coil ignition and capacitor discharge ignition led to electronic ignition (semiconductor ignition [SI]) and fully electronic ignition (breakerless semiconductor ignition [BSI]; see Appendix, Ignition systems). In the case of SI, an ignition distributor was still present; however, in BSI, this was replaced by a multi-spark coil system. The first application of BSI was found in the Citroën Visa subcompact, introduced in the fall of 1978. The Visa still employed a conventional downdraft carburetor.

In the latter half of the 1970s, high-priced cars employed two separate electronically controlled systems, namely, ignition and fuel injection. If these individual systems are combined, the result is coordinated control of engine processes, also known as electronic engine management. These systems are more flexible than individual systems and can be expanded by fitting additional subsystems such as closed-loop oxygen sensors, idle control, and knock sensing and control.

In 1979, Bosch introduced such an engine management system under the designation "Motronic." For the first time, a microprocessor (i.e., software) was installed as an integral part of a motor vehicle. Digital electronics had replaced analog technology. Motronic could alter dwell angle, ignition timing, fuel mixture, and injection timing in accordance with complex curves and maps, and could do so with much higher accuracy than any mechanical system. Fuel economy and exhaust emissions improved, as did engine response and power, with higher compression ratios and leaner mixtures. Moreover, the safety margin from the knock limit could be reduced, making the engine largely independent of fuel quality.

The first car to be fitted with Motronic engine management was the 1979 BMW 732i (Figure 171). Further engine management systems soon followed: Weber in 1980, Nissan in 1981, Renix in 1984, and Volkswagen (Digifant) in 1987.

In the United States, the combination of BSI and electronically controlled throttle body injection, employing only a single fuel injector, was introduced around 1985 and represented a more cost-effective solution than individual injectors for each cylinder. Because fuel droplets are deposited on the manifold walls to varying degrees, making cylinder-to-cylinder mixture distribution less

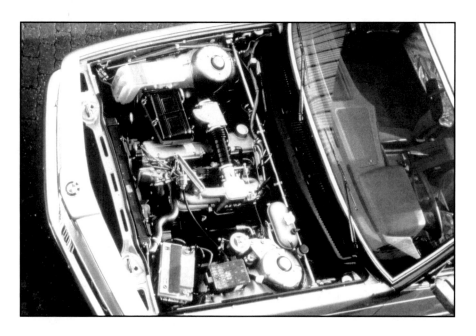

Figure 171. BMW 732i, 1979. Bosch L-Jetronic fuel injection system in the engine bay of the flagship sedan of BMW, with intake manifold below the valve cover (driver side), above that the throttle and throttle switch, and, on the other side of the bellows, the air flow meter and intake air temperature sensor. At 16.2 liters/100 km (14.5 mpg), the 197-hp injected engine achieved slightly better fuel economy than its less-powerful carbureted predecessor, the 184-hp 730 (16.5 liters/100 km, or 14.3 mpg). Six-cylinder engine, 3210 cc, 197 hp (145 kW).

uniform, this system is inferior to multi-point injection. Bosch offered such a central injection system in 1983, followed by Honda in 1988.

A further development is represented by the Saab Trionic engine management system, introduced in 1992 (Figure 172). At its core is a 32-bit processor that can perform two million calculations per second, monitoring and controlling ignition, fuel injection, and turbocharger boost pressure. New features included two-channel injectors and measurement of the electrical conductivity of combustion gases (ionization) through the spark plug, making the knock sensor superfluous.

Trionic also incorporates direct or capacitor ignition, introduced by Saab in 1987. In this system, voltage is first transformed to 400 V and, after storage in a capacitor, is again transformed to the spark plug voltage of 30 to 40 kV (instead of the 25 kV of conventional coil ignition). Because this system requires fewer coil windings, each spark plug is given its own ignition coil, mounted directly on the spark plug. The advantages of this design are faster

Figure 172. Saab 9000 CS 2.3 Turbo, 1992. In 1993, Saab introduced its own engine management system for its four-cylinder 2.3-liter turbocharged engine, combining the functions of (indirect) fuel injection, ignition (Saab direct ignition), and boost control in a central monitoring and control unit. Other features include two-channel fuel injectors, capacitive direct ignition with multiple sparks per cylinder firing, and using the spark plugs as ionization detectors to sense whether the mixture is rich or lean. Double overhead camshafts, four valves per cylinder, four-cylinder engine, 2290 cc, 200 hp (147 kW).

voltage rise, more powerful spark, multiple spark (up to 50 firings per cycle) for starting, and elimination of voltage losses by deleting the ignition distributor, spark plug leads, and separate ignition coil.

The multi-point or throttle body injection systems previously mentioned employ indirect, or manifold, injection, in which fuel injection nozzles are located either just behind the engine intake valves or near the throttle. The auto industry is seeking further reductions in fuel consumption and exhaust emissions by employing direct injection, in which injectors deliver fuel directly into the combustion chambers. The industry hopes to achieve 20 percent cuts in both fuel consumption and carbon dioxide emissions, coupled with a power increase of 10 percent. Balanced against these gains are higher design and development costs. Fuel must be supplied by a high-pressure pump at 50 to 120 bar (220 to 530 psi); manifold injection needs no more than 5 bar

(22 psi). To reduce nitrous oxides resulting from lean combustion, the system will require installation of an additional NO_x catalyst. However, this predicates sulfur-free fuel, which was introduced (to the German market) by the oil industry only in the summer of 2000.

As a result, other nations have taken the lead in direct-injection development, despite the historical leadership of Germany in this field (1924, MAN four-stroke diesel truck; 1951, Goliath and Gutbrod gasoline two-stroke; 1954, Mercedes and Borgward gasoline four-strokes; 1990, Audi diesel four-stroke passenger car, Figure 173). In 1995, Mitsubishi introduced its Carisma, powered by a direct-injection gasoline engine (Figure 174), sold in Germany as of 1997. This was followed by the Toyota Avensis D4 in 1996, Volvo S/V 40 (with a Mitsubishi engine) in 1998, and the Renault Mégane coupe and convertible in 1999, with the firm's own IDE engine. For 2000, Volkswagen introduced the Lupo FSI powered by direct-injection gasoline engines.

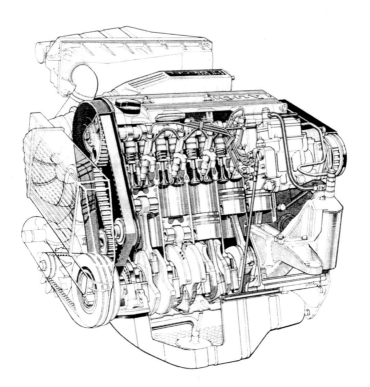

Figure 173. Audi diesel engine, 1990. The first German direct-injection diesel engine with overhead cams, distributor injection pump, and bowl-in-piston combustion chambers. It exhibits a high degree of technical complexity: electronic diesel engine management, dual-spring injection nozzle holders for pre- and main injection, turbocharging, charge air intercooling, and boost control. Fuel economy is approximately 15 percent better than a comparable prechamber engine. Five-cylinder diesel engine, 2460 cc, 115 hp (85 kW).

Figure 174. Mitsubishi Carisma GDI, 1995. A midrange model powered by a gasoline engine, with fuel injection directly into the combustion chambers. The engine employs two very different operating programs. In the low part-throttle regime, it is a lean-burning stratified charge engine. In the remainder of its range, it is a performance-oriented fuel injected engine. Fuel economy is better than comparable gasoline engines with indirect (manifold) injection, but not as good as direct-injection diesel engines. Four-cylinder engine, 1834 cc, 125 hp (92 kW).

All of the these automobiles employ electronically controlled fuel injection systems and catalytic converters. These features are almost universal today, but when introduced to the American market of the 1970s and a decade later in Europe, they were met with stubborn resistance. Catalytic converters operate only with unleaded gasoline and are effective only when coupled with oxygen sensors. These measure the oxygen content of the exhaust stream, permitting electronic correction of the air-fuel ratio.

The realization that toxic exhaust gases require post-treatment is by no means a modern achievement. As early as 1910, chemist and journalist Walter Ostwald recommended that "the exhaust gases undergo catalytic combustion...they be mixed with a suitable quantity of auxiliary air and passed over a substance (e.g., copper oxide, 'purple ore' [iron copper sulfide])...the heat of the exhaust gases (300–350°C [572–702°F]) quickly raises them to an appropriate temperature, to complete the combustion process...The difficulty of this principle lies in the fact that contact is not achieved immediately upon starting the engine, but rather only after some delay." (*Autler-Chemie*, Berlin, 1910). Obviously, Ostwald was well aware of cold-start difficulties.

Today, electronics may be found in any and all automobile subassemblies. It is beyond the scope of this book to examine these in more detail, beyond brief mention. In the drivetrain department, applications include engine management, transmission control, diesel governors, and many others. In active safety systems, they are found in anti-lock brakes and traction control, vehicle dynamic control, and tire pressure monitoring, to name but a few. Electronics contribute to passive safety as triggers for airbags, belt tensioners, automatic roll bars, radar distance sensors, and related systems. Occupant comfort is enhanced by electronic heating, ventilation, and air conditioning controls, seat adjustment with memory, central locking, and similar features. Electronics are at the heart of radio and entertainment systems, on-board computers, navigation systems, and an ever-expanding plethora of in-car communication and information systems.

The increasing application of individual subsystems led to a jungle of cables, positioners, control units, connectors, and sensors. An automobile of 1949 needed a wiring harness consisting of perhaps 35 meters (38.3 yd) of wire and 16 connectors; however, by 1965, this had grown to 150 meters (164 yd) and 200 connectors, and by 1990 expanded to 2,000 meters (2,187 yd) and 1,500 connectors. Moreover, this added to vehicle weight and required extensive (hand) assembly work. Worse, the systems were interlocked; for example, traction control could modulate engine torque by interfacing with the engine management system.

To achieve greater reliability, lower weight, reduced space requirements, and automated assembly, since the 1980s both General Motors and Ford/USA have made use of time-multiplexed systems, familiar since the 1930s in communications technology. These consist of a local area network, transferred to the automobile, with two wires: one carries current from the car battery to various substations, while the other carries commands and replies between the central station and substations. Also known in Germany as a CAN bus (Controller Area Network), such a network permits complete data exchange among various controllers, using only a single conductor. The first two vehicles to employ multiplex wiring were most likely the Cadillac Allanté and Lincoln Continental, both of 1987, followed in Germany by the 1990 BMW 850i.

Power requirements will increase, particularly after introduction of various "by wire" systems, and vehicle electrical systems will have to be modified accordingly. The original Volkswagen Beetle needed no more than 140 W at 6V, whereas the New Beetle's 12V system requires up to 2 kW. The coming years will probably see system voltages rise to 42V.

Meanwhile, the spatial juxtaposition of mechanical and electronic components, and their accompanying software, has turned the automobile into a mechatronic systems carrier. A key characteristic of these mechatronic systems is their ever-increasing miniaturization and attendant advantages: expanded functionality, such as electronically controlled throttle (E-Gas) with speed control (cruise control); reduced assembly requirements, for example, delivery by a supplier of an automatic transmission complete with control unit, which once had to be mounted separately under the seat; and generally reduced space requirements.

Examples for applied networking of mechanics, hydraulics, electricals, electronics, and software include electromagnetic (engine) valve control, common rail injection systems (see Appendix, Diesel injection), ABS hydraulic unit with electronic stability program (ESP) and vehicle interval (following distance) control, and electronic control of convertible tops, windows, and roll bars that deploy in the event of a rollover accident. Common to all of these examples is data acquisition by means of micromechanical sensors that measure physical quantities and generate electronic control signals.

Without electronics, the automobile—similar to other capital and consumer goods—has no future. In 1992, electronics made up approximately 15 percent of the value of an automobile. For 2010, this is expected to rise to 25 percent. According to a study conducted by Intel and McKinsey, in 1990 the average purchase price of electronic components per car amounted to approximately \$1,200 in the United States, \$950 in Japan, and \$600 in Europe. Year by year, this portion will rise, and, perhaps sometime in the future, Europe will close the gap. However, North America is likely to retain its lead, the result of American emissions, safety, and fuel economy regulations.

Auto Production in Other Countries

If scientific awards are any indication, scientists and engineers working in Asia rarely make groundbreaking discoveries. The magazine *Bild der Wissenschaft*, Vol. 4/1993, stated that "The three Nobel Prizes [that went to Japan] in the past 25 years are a meager harvest compared to the United States, 77 of whose scientists were so honored in the same time frame." This, of course, assumes that the Nobel Prize is an appropriate yardstick for gauging the level of scientific research.

The reality that basic scientific discoveries have little relation to establishing an effective automotive industry has since been proven by Japan and South Korea. In 1967, Korea was a backwater of the automotive world; however, 30 years later, Korean car makers produced 2.85 million units (cars and commercial vehicles), putting them in fifth place behind France, Germany, Japan, and the United States—and three years ahead of schedule. However, the Southeast Asian economic crisis of 1997–1999 temporarily put the brakes on Korea's scramble up the charts; production dropped to 1.65 million units in 1998, relegating Korea to seventh place.

The Korean firms Hyundai, Kia, Daewoo, SsangYong, Samsung, and commercial vehicle maker Asia did not by any means produce better vehicles than their competition. However, similar to the Japanese before them, they took advantage of favorable economic and political conditions: low wages, aggressive government economic policy, cunning import restrictions in terms of both tariffs and psychological considerations, high pent-up demand in neighboring Asian countries including China, and a demand for cheap cars in the U.S. market around 1989 after price hikes for Japanese cars. Added to these were the industriousness and (initially) modest wage demands of Korean workers.

Hyundai, as the largest Korean manufacturer, began by assembling the British Ford Cortina in 1968. This was replaced by its own successful Pony model (Figure 175) in 1975. The Pony was based on a Mitsubishi design but had its own unique body, drawn by Giorgio Giugiaro/Ital Design. Ital Design also designed bodywork for Alfa Romeo, Fiat, Maserati, Audi, and other clients, including Volkswagen. The Pony had more than a passing resemblance to the Volkswagen Passat (Dasher in the U.S. market). In 1988, Hyundai reached fifth place in American new-car registrations with 267,000 units. In 1991, Hyundai produced its four millionth vehicle. German operations began in 1991.

Figure 175. Hyundai Pony, 1975. Lower midrange model with clean body lines by Giorgio Giugiaro/Ital Design, similar to those of the Volkswagen Passat (Dasher in its earlier American incarnation). Running gear by Mitsubishi, still employing live and beam axles on semi-elliptic springs and rear-wheel drive. Four-cylinder engine, 1238 cc, 82 hp (approximately 60 kW).

Kia, the second largest Korean car maker, began offering light commercial vehicles in 1962. It began license production of Mazda designs the following year. In 1976, Kia commenced production of heavy trucks and buses under the Asia brand. Its first independent passenger car design was the 1993 Sephia. Kia, operating in Germany since 1993, hired Karmann of Osnabrück to assemble its Sportage sport utility vehicle (Figure 176) between April 1995 and July 1998.

Figure 176. Kia Sportage, 1995. Korean four-wheel drive all-terrain and leisure vehicle. This particular example was built in Germany, intended as the first step in ongoing research and development efforts between Kia and Karmann. German production ceased in July 1998, after 25,960 units were built for the domestic market. Four-cylinder engine, 1998 cc, 95 hp (70 kW) or 128 hp (94 kW).

For almost 30 years, Daewoo, a conglomerate founded in 1967, drew on various models by Opel (LeMans, Espero, and Nexia), Suzuki (Tico), and Honda (Arcadia). In 1997, Daewoo brought out its own independent designs, created under the leadership of former Porsche chief engineer Ulrich Bez. The smartest of these was the subcompact Matiz, with a Giugiaro body design. Daewoo has been in the German market since 1995.

SsangYong, a highly diversified corporation, entered the automotive scene in 1969, building Kaiser Jeeps under license. In 1974, SsangYong began its cooperative ventures with American Motors Corporation, and later license-built Isuzu off-road station wagons, some of which were powered by Mercedes engines. SsangYong has been active in the German market since 1996.

Samsung, another diversified concern, began truck production in 1994. Production of Nissan cars under license commenced the following year, interrupted by a production halt of several months. Samsung made a second attempt at car production in 1999, in a brand-new, multi-billion dollar plant.

For years, it seemed that the sky was the limit for Korean auto makers. In all seriousness, they believed they could engage in an all-Asian economic policy, completely separated from the West, free of economic downturns and with guaranteed growth. They increased production capacity to 5 or 6 million vehicles annually, created transplant assembly facilities in (among others) Canada (Hyundai), Iran (Kia), and Spain (SsangYong), and formed partnerships or took over firms such as FSO (Poland), Avia (Czech Republic), and Oltcit (Rumania). With Hyundai and Daewoo's stated predictions that they would be among the ten largest car makers in the world by 2000, the Koreans particularly irritated the management boards of their Western and Japanese competitors.

However, the same laws of economics apply throughout the world, apparently without regard to individual regions. The trigger for the current demise of the Korean economy was the monetary crisis that began in 1997. This exposed weaknesses throughout Korean industry but in particular the automotive sector, where an expansionist strategy had amassed mountains of debt despite cross-subsidies from other branches of industry. With the collapse of Kia in 1998 and Daewoo in 2000, it became obvious that the industrial and commercial base of the country was in serious need of reform. Also, five auto makers are too many for a nation of approximately 43 million inhabitants (1992 population). Another weakness of the Korean auto industry was its overly long dependence on foreign licenses.

At present, Korean restructuring is in full swing. A snapshot at the turn of the millennium provides the following image:

In 1998, Hyundai took over a bankrupt Kia Motors, including its commercial vehicle division, Asia. DaimlerChrysler in turn took a 10 percent share of Hyundai in June 2000. Daewoo devoured SsangYong at the end of 1997 but soon faced difficulties of its own. After fruitless negotiations with General Motors, its creditors did something unusual in the auto industry—they placed Daewoo up for bids. Ford won the sole right to negotiate for Daewoo but refrained from taking over the firm because of its high liabilities. In the spring of 2000, Samsung Motors Inc. came under Renault control.

Does this mean that the South Korean auto industry has been domesticated? Perhaps, but at the least, it has been swept up by the general wave of globalization. By any rational and objective view, no other course would have been possible. However, it would be rather short-sighted to underestimate the potential of South Korean industry. The Japanese emerged from every crisis stronger than ever; the same may be expected of the Koreans.

In 1999, of 3.8 million new vehicle registrations in Germany, imports achieved a market share of 33.7 percent. Japan accounted for 11 percent, France had a 10.8 percent share, Italy 3.9 percent, and 1.34 percent (51,114 units) came from South Korea.

The market share of the Malaysian auto industry is minuscule compared to these numbers. Malaysia is the third Far Eastern competitor on the German market. Of the two or three auto firms operating in Malaysia, only Perusahaan Otomobil Nasional Berhad, better known as Proton, plays any significant role. On government initiative, the firm was founded in 1983–1985, with technical assistance from Mitsubishi. Proton has been supplying the British market since 1989 and Germany since 1995. In 1998, it managed slightly fewer than 2,500 new passenger-car registrations in Germany. Even today, the influence of the Mitsubishi Colt, Lancer, and Galant models is unmistakable.

Japan's dominant position in technology and manufacturing methods encompasses the entire Far Eastern region, including Taiwan and (to a limited extent) the People's Republic of China. China's auto history began in 1955 with license-built Soviet trucks, buses, and tractors. At first, passenger cars for the Party and various government agencies were imported from the U.S.S.R. In 1960, China started its own production of the official Hongki (Red Flag) limousine and the midrange Feng Huang (Phoenix) models. The Cultural Revolution of 1965–1976 stifled any embryonic efforts toward independent technical development. Thus, in the 1980s, the Chinese auto industry had to rely on imports and foreign aid.

Until around 1990, China's growth in motor vehicle traffic concentrated on commercial vehicles and buses. Originally, passenger cars were intended to serve only the needs of party functionaries or as taxicabs. The Eighth Five-Year Plan of 1991–1995, however, made it easier for private individuals to obtain motor vehicles. The result was that Chinese motorization took a great leap forward, as it were. Likewise, it helped China to its current status as the seventh largest economic power in the world.

In the commercial vehicle sector, the main beneficiary of this growth was primarily MAN subsidiary Steyr-Daimler-Puch, operating in China since 1984. Among car makers, AMC-Chrysler-Jeep (1982), the Volkswagen group with its Santana, Jetta, Golf (1983), and Audi 100 (1988) models, Daihatsu, Subaru, and Suzuki (all 1986), and PSA with Peugeot (1988) and Citroën (1992) all had Chinese operations.

Fortunately for the German auto industry, a review of the Chinese situation around 1994 showed Volkswagen in the lead at 40 percent market share, with the Audi 100, renamed "Red Flag," holding pride of place as the vehicle of choice for notables. From the Chinese viewpoint, however, the auto boom of the early 1990s often generated chaotic conditions. High customs duties gave rise to organizations involved in smuggling parts and entire cars. Auto plants and dealerships sprouted in huge numbers, while a fragmented industry operated in an uneconomical and labor-intensive manner. In 1994, to regain control of the development of the auto market, the government instituted various restrictions, prohibited any further joint ventures in 1995, and in 1994 invited all interested foreign manufacturers to take part in the "Family Car Project."

The intention was to develop a state-of-the-art family car expressly for Chinese conditions, following specifications set forth by the Ministry of Machine-Building Industry. These included compact exterior dimensions, the ability to transport four persons in comfort, minimal fuel consumption, and a price of less than $10,000. Of the 24 European, American, Japanese, and Korean firms that took part, only Porsche and Mercedes fulfilled Beijing's requirements. Both demonstrated subcompact cars developed specifically for the Chinese market, with variable bodywork. Chrysler took a project intended for the South American market, the Composite Concept Vehicle, and in short order renamed it the China Concept Vehicle. Other manufacturers fell back on existing models or their planned successors, which proved to be a wise strategy. The Chinese government was not interested in a long-term presence of foreign car makers but rather their know-how.

Parallels to the original Volkswagen Beetle story were obvious. Then as now, a regime without any democratic legitimacy had bypassed the industry's usual process to develop its own people's car. The NSDAP (the Nazi Party) and its subsidiaries arranged to have at least part of the design, development, and prototyping costs of the Beetle advanced by KdF "savers"—ordinary citizens who set aside regular amounts in a "savings book" toward purchase of a car. On the other hand, the Chinese Communist Party's own Ministry of Machine-Building Industry had the contestants hand over plans and prototypes at no cost, presumably intending to build such a car on their own.

To date, that has not happened. Quite the contrary, because China could not make the connection to world-class auto manufacturing standards on its own, General Motors signed a cooperative agreement with China in 1997 for production of its Opel Astra or Vectra models. The intention was to break the dominance of the Volkswagen group, which had grown to 60 percent of the market. However, what are actually being built are Buick and Chevrolet models.

After the withdrawal of Peugeot in 1997, Volkswagen/Audi (Figure 177), Daihatsu/Toyota, Suzuki, and Citroën accounted for approximately 90 percent of Chinese production, with a volume of 523,760 (1999)—rather modest for a country with a population of 1.2 billion. Not surprisingly, Western and Japanese auto makers view China as a highly promising export market for the future.

India, too, with more than 900 million inhabitants and an annual production of only 484,328 passenger cars and station wagons (1999), has its automotive history still ahead of it. Into the 1980s, the Indian auto industry usually built older European models for which it had obtained licenses and tooling. The Premier was the old Fiat 1100 and 124 types, the Hindustan was the reincarnated Morris Oxford and Vauxhall Victor, the Mahindra was an Indian-built Jeep, and the Bajaj was the Tempo three-wheeled delivery vehicle once popular in postwar Germany.

The Maruti firm was founded on plans by Indira Gandhi's son Sandjay to build an Indian "Volkswagen." In 1982, Suzuki bought into the firm and brought modern technology to India in the form of its Alto, Swift, and Minibus models. Sandjay Gandhi, who died in an aviation accident in 1980, did not live to see at least part of his wishes come to fruition. Today, Maruti is India's largest passenger car manufacturer.

Figure 177. Shanghai-Volkswagen Santana LX/2000, circa 1983-1995. Similar to Volkswagen do Brasil and Volkswagen Argentina, Shanghai-Volkswagen Automotive Corporation has been building the Volkswagen Santana (known as the Quantum in the U.S. market) since the early 1980s. The Santana entered the German market in 1981 as a notchback version of the Passat but was never really accepted by the public. In Brazil, Argentina, and China, it appeared under various names and underwent further development. Four-cylinder engine, 1588 cc, 75 hp (55 kW) or 1781 cc, 98 hp (72 kW).

After the once rigorously protected market was fully opened in 1993, car makers great and small announced cooperative efforts with firms in India: Daewoo (joint venture, 1994); Honda (1995); Hyundai (1996); and Fiat. The Tata Engineering and Locomotive Company (Telco) is planning exports to Europe. One obstacle continues to be posed by the lack of a middle class in India. Government requirements that new auto plants produce cars with at least 50 percent local content after three years of operation have caused foreign firms to tread carefully when considering investment in India.

Gone from the scene are several firms from the former Soviet Bloc, including the Trabant and Wartburg nameplates. East German auto makers were not the only branch of industry to suffer with political and ideological handicaps that were not overcome until the collapse of the German Democratic Republic in 1989. In the early postwar years, destruction, looting, dismantling, and expropriation hampered reconstruction and production. Lack of investment credits, supplier industries, and steel plants impaired planning, production, and ability to compete. The Communist ideology, rooted in the concept of collective ownership, thwarted individual motorized traffic, as witnessed by the backward nature of East German vehicle technology, the abysmal condition of the roads, and the thinly spread network of repair facilities.

Thus, the Wartburg and especially the Trabant (Figure 178) remained automotive fossils that could not hope to approach the standards of Western automobiles. Record-setting statistics that include total production of more than 3 million Trabants, a 33-year production run, and delivery waiting times of up to 15 years can be readily dismissed as symptomatic of the East German system. In the harsh climate of free-market competition, the Trabi outlived "Socialist Realism" by only 17 months.

Figure 178. Trabant P 601, circa 1964. In 1958, the VEB Automobilwerke Sachsenring ("VEB" stands for "Volkseigener Betrieb" or "People's Own Company" in the former East Germany) began production of its subcompact Trabant, a car built in the DKW tradition: front-wheel drive, two-cylinder two-stroke engine, and non-steel body. Although the Trabant has a steel unit structure, this is covered by synthetic exterior panels. These consist of 80 to 90 layers of cotton waste crepe in a phenolic resin matrix, formed by steel rollers and heated presses. Two-cylinder two-stroke engine, 594 cc, 23 hp (17 kW).

With the help of Fiat and Renault, other auto manufacturers of the erstwhile Comecon states had managed to make the connection to Western automotive technology. The Lada/Zhiguli plant, built with the help of Fiat in Togliattigrad, U.S.S.R. (now in Russia), produced 730,000 cars in 1980 (Figure 179), and even 18 years later still managed almost 600,000 units. Polski Fiat likewise manufactured approximately 300,000 cars in 1998, compared to almost 200,000 in 1982. Dacia, a Renault licensee since 1968, was the Rumanian market leader with 106,000 cars and pickups built in 1998. The following year, control of Dacia passed to Renault.

Figure 179. Lada 1200/Zhiguli VAZ 2101, circa 1976. The VAZ Auto Works, near the Zhiguli Hills outside Togliattigrad, Russia, continues to export its Zhiguli VAZ 2101 under the designation "Lada." The Lada is a copy of the 1966 Fiat 124, built in the former Soviet Union since 1970 and exported since 1972. Compared to the Fiat original, the Lada features, among other things, its own engine design, thicker steel for the bodywork, and increased ground clearance. Four-cylinder engine, 1189 cc, 60 hp (44 kW).

The situation in all of the countries considered here is representative of all other nations and regions outside Western Europe and the United States. The lessons to be learned are twofold. First, the influence of the two largest American concerns, General Motors and Ford, and increasingly the influence of Western European and Japanese firms, is readily apparent in all of the young auto-building nations. Second, with the exception of Japan, none of these nascent auto nations has made any significant contribution to automotive technology or production methods.

Commercial Vehicles and Buses, 1945 to 2000

The immediate postwar years were occupied with removing the rubble of war and rebuilding ruined cities and their transportation infrastructure. These tasks called for trucks and delivery vehicles. In at least one case (Büssing), truck production resumed, with permission of the occupying Allied forces, even before the German Wehrmacht's unconditional surrender on May 7–8, 1945. In the remaining months of 1945, Tempo, Mercedes, Ford, Büssing, MAN, and Kaelble even managed to produce 5,315 vehicles—an incredible achievement in view of bombed-out factory buildings, scarce raw materials, worn-out machinery, and workers and employees killed at the front or still in prisoner-of-war camps.

In 1949, Volkswagen introduced its Transporter, powered by an air-cooled engine at the rear (Figure 180). In sheer numbers, it far outsold all competitors and, in the same way as the Beetle, soon became a success in export markets. Its success prompted Fiat/OM and Chevrolet to build similar rear-engined delivery trucks.

Figure 180. Volkswagen Type 2 (Transporter), 1955–1959. Reliability, ride comfort, and economy of operation were the most significant attributes of the Volkswagen Transporter, which began production in 1950. It was available as a delivery van, station wagon, pickup, and eight-seat bus. With almost 7 million examples built, the Type 2 was a successful commercial vehicle. However, because its rear-engine layout eventually proved to be unsuitable for further development, Volkswagen has been building front-engine, front-drive delivery vehicles since 1990. Four-cylinder engine, 1192 cc, 30 hp (22 kW).

A year earlier (1948), DKW, Tempo, and Wendax had introduced front-drive delivery trucks with forward control and semi-forward control, patterned on Citroën's TU of 1939 (Figure 116). These were followed by Lloyd (1952), Goliath (1953), Alfa Romeo (1954), Renault (1958), Barkas (1961), Lancia (1964), and Fiat (1969), all except Lloyd with water-cooled engines. In the process, delivery vans acquired their own unique layout and appearance.

The excellent market position enjoyed by German front-drive vans faded away after the Borgward Group (Goliath, Lloyd) left the market in 1961, and DKW terminated production of its express truck in 1962. The Tempo Matador remained, and in the following years and because of changing ownership, it would be marketed as a Hanomag, Hanomag-Henschel, and Mercedes. Between 1977, when production ended, and the introduction of Mercedes' Spanish-built 100 D in 1987 and the Volkswagen T 4 in 1990, there was a market void that Fiat sought to fill with its Ducato (1982, Figure 181). The front-drive Ducato not only mounted the transverse engine and transmission ahead of the front axle; the head of the vehicle also housed the fuel tank, spare tire, and battery.

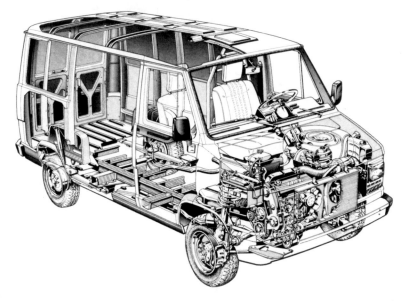

Figure 181. Fiat Ducato 14 Turbodiesel, 1987. The Ducato was designed to meet the demands of commerce, with all mechanicals housed in the head of the vehicle, simple rear suspension, and flat load floor with no sill or cross member separating it from the cab. This simplifies engineering changes in wheelbase, track, axles, ground clearance, and overall length, giving a multitude of possible chassis and bodies. Transverse engine, front-wheel drive. Four-cylinder diesel, 2455 cc, 92 hp (68 kW).

Ready-to-drive cabs and chassis with this layout, first offered by Latil in 1907 (Figure 182) and later by DKW in 1949 and Lloyd in 1952, are ideally suited to a variety of special bodies and structures, especially if track, wheelbase, axles, or ground clearance require modification.

The next class above delivery vans, with payloads of 1.5 to 3.5 tons, was at first dominated by hooded units incorporating American styling elements: emphasis on horizontal lines, with a broad grille, "alligator" hood, and chrome trim. These features were adopted by Hanomag in 1950, by Ford in 1951, by Opel in 1952 (Figure 183), and by Borgward in 1954. Next, the forward control layout achieved dominance beginning with Ostner in 1955, followed by Mercedes (1956), Borgward and Faun (1957), and Hanomag (1958). This design had been common in the English-speaking world and southern Europe since the 1930s, and it was first seen in heavy German trucks from the mid-1950s, led by Büssing's 1952 entry.

Figure 182. Latil 1.75-ton truck, 1907. An interesting vehicle by George Latil, used to transport injured horses. The four-cylinder engine is installed transversely ahead of the front axle. The radiator is separate from the engine. The vehicle has front-wheel drive and a tilting truck box (which may be replaced by other specialized structures) attached to pivots on the frame, and a ramp leading to the box. Latil's principle lay forgotten for decades until it was revived for small car transporters. Development of a similar concept, the Ruthmann-Kolli lift system, began in 1949. These tilt-bed concepts are possible only with front-wheel drive.

Figure 183. Opel Blitz, 1952–1960. The 1.75-ton succeeded the 1.5-ton Blitz, built between 1938 and 1942, and again after the war from 1946 to 1952. Following the example of American trucks, it too was given a comfortable cab with flowing exterior lines. As always, Opel offered only gasoline engines in its commercial vehicles. Six-cylinder engine, 2473 cc, 58 hp (43 kW) and 62 hp (46 kW), as of 1955.

In Germany, declining importance of the railroads prompted regulatory agencies to introduce drastic length and weight limits for trucks in 1956 and 1958. The immediate consequences of these so-called "Seebohm" laws, which were later relaxed, were layoffs and cutbacks, and separate development of domestic and export models, but eventually the adoption of internationally accepted standards: lightweight design, forward control, fifth wheel, and more power. Although early tests in the 1950s often ended in failure, eventually most manufacturers fitted their engines with exhaust-driven turbochargers to meet the new specific output requirement of 6 hp/ton, which took effect in 1960. Intercoolers, introduced to Europe by DAF in 1973, permitted even greater power increases. The shift from prechamber engines to more economical direct-injection diesels, as practiced by MAN since 1924, was undertaken by Krupp (using Cummins engines) in 1963, Mercedes in 1964, and Büssing in 1965, followed by other manufacturers.

In the early 1960s, a fresh, "European" forward control design replaced the less self-confident styling of the 1950s. Henschel was first in 1961 (Figure 184), followed by Magirus (1963), Büssing (1965), and Steyr (1968), all with cabs designed by Louis L. Lepoix.

Figure 184. Henschel HS 14 TL, 1961–1965. At the 1961 Frankfurt Auto Show, Henschel showed its revamped truck line, characterized above all by the modern lines and interior appointments of their cabs. These were the work of French industrial designer Louis Lucien Lepoix (1918–1998). More than any other designer, Lepoix put his imprint on an entire generation of commercial vehicles. Six-cylinder diesel, 11,045 cc, 180 hp (132 kW).

This represented a change in the German commercial vehicle industry. After a long hiatus, it was again commissioning independent designers to style its trucks. As early as 1914, poster and graphic artist Ernst Neumann-Neander had drawn a truck forward section for Büssing (Figure 185), of which several styling elements such as radiator, hood, and "chin spoiler" including corporate script were incorporated in production models. This may have been the world's first instance of commercial vehicle design being contracted to an artist outside the truck industry.

Figure 185. Büssing front end, 1914. Writing in the 1914 German Werkbund year-book, Ernst Neumann-Neander ("N²") reported that "The pressure of competition...is beginning to shake the truck manufacturer. Now he is forced to seek good shapes for these vehicles as well." Certain of N²'s styling elements would resurface in later Büssing trucks.

In 1965, for one Büssing truck model, the Lepoix unit cab could be dropped so low (thanks to an underfloor engine) that the cab floor was only 57 cm (22.4 in.) above the road surface. Josef Bertsch, engineer in charge of the project, took the next logical step and placed the cab below the cargo bed. The result was the Büssing Supercargo deck truck—also of 1965 (Figure 186)—with its 10-meter (32.8-ft) cargo bed in an overall length of 10.3 meters (33.8 ft), which provided a load surface almost as large as the footprint of the vehicle. Despite later attempts by Strick Corporation of Pennsylvania in 1978, again using a Bertsch design, and Steinwinter/Drögmöller/Schmitz in 1983, the cargo deck principle has not been pursued, presumably because it is too unusual and too futuristic.

In May 1966, the first container carried by a regularly scheduled shipping line was lifted from the first dedicated container ship to dock in Bremen. For Germany, as in many other countries,

Figure 186. Büssing Supercargo 22-150 deck truck, 1965. By moving the engine and cab below the load floor, it is possible to achieve even greater load surface and space than in conventional forward-control designs. The deck truck can be configured as a volume or container transporter, or as a fifth-wheel tractor with an additional container above the cab and engine. With the front bumper against a loading dock, overhead loading is possible. Six-cylinder diesel, 7416 cc, 150 hp (110 kW).

this began a new chapter in transportation. Special container vehicles were developed for road, rail, and water transport, necessitating basic, underlying changes to vehicle layout and engineering, as well as transportation logistics.

Before the arrival of standardized containers, which American forces had employed in great numbers during World War II, various national railway agencies had used their own container designs but without any uniform design standard. In 1834, a serious attempt to eliminate the separation of cargo and container was represented by the shift from carriage bodies to railway cars on the Paris-Orléans rail line. Nineteen years earlier, in Nymphenburg near Munich, Joseph Ritter von Baader had demonstrated a functional model of a roadable railway train, a hybrid that could switch between two modes of transport.

In the 1950s, bus technology finally divorced itself of its truck heritage, with its body-on-frame construction, rigid axles, longitudinal leaf springs, and front engine location. In terms of ride and ease of operation, the passenger car would thenceforth serve as a role model. For interior equipment and weight reduction, modern buses drew their inspiration from aircraft.

Development of unit-bodied buses, interrupted by the war, was taken up again, accelerated by company policy considerations and economic factors. For one thing, Daimler-Benz AG no longer provided chassis to the roughly 50 bus body manufacturers because the firm was no longer interested in supplying its own competition. Furthermore, lightweight design could not be achieved with conventional body-on-frame construction, with a payload ratio of about 0.5. Steel unit bodywork raised this to 0.77 and aluminum alloys to approximately 0.9 (1952).

More from necessity than anything else, independent bus body manufacturers took the lead in bus design because the major truck manufacturers, with an eye toward recovering capital investment and sharing components across different lines, resisted adoption of the unit-body concept. In 1951, all-steel unit-bodied forward control buses came on the market, powered by diesel engines at the rear. These included designs by Otto Kässbohrer and Theodor Pekol, and built by Kässbohrer/Setra; Berliet in France; and Krauss-Maffei, NWF, and Drauz. The last three were based on designs by Henrich Focke and differed from one another only in minor details. Also in 1951, Orenstein & Koppel built a unit-bodied all-steel double-decker on a Büssing chassis. In 1952, Klatte introduced a unit-bodied aluminum single-decker.

Engines were still installed longitudinally. Although the transverse engine/front control layout, with the powerplant mounted below the last row of seats, offered approximately 20 percent greater seating capacity, the bus manufacturers listed here, as well as later makers, rarely took advantage of this configuration.

The German bus industry was motivated to adopt this layout by Munich armaments maker Krauss-Maffei, which had never built a bus. The immediate postwar years saw an above-average demand for buses, and Krauss-Maffei had a good number of Maybach gasoline engines on hand, once intended for the 8-ton halftrack. In November 1945, a U.S. Army major, apparently with industry experience from civilian life, cranked up Krauss-Maffei's bus production but insisted on a forward-control design with a longitudinal engine at the rear (Figure 187).

Figure 187. Krauss-Maffei KMO 130, 1947. By order of a U.S. Army major serving with the occupation forces, Munich's Krauss-Maffei mounted an engine at the back of an upcoming bus project. The vehicle was ready for trials in January 1946. Without a body and using a temporary driver's seat, the American officer drove it to Army headquarters in Garmisch to obtain production permission. Soon thereafter, Krauss-Maffei began its bus program, in cooperation with the Rathgeber works, which built the bodies. Six-cylinder Maybach gasoline engine, 6191 cc, 135 hp (99 kW).

Other firms, however, saw great advantages in underfloor engines. In 1949, Büssing introduced a tram bus with a newly designed underfloor powerplant, to be followed by Renault and FBW in the same year, Berliet in 1951, Mowag and Henschel in 1953, and Mercedes and MAN in 1957.

In 1953, Graaf and Orion introduced four-wheel independent suspension via coil springs, following Pekol's designs. Wilson and ZF offered transmissions that could be shifted under load, Voith offered hydromechanical transmissions, and Klatte in 1951–1952 offered hydrostatic transmissions on an experimental basis. Independent bus heating systems appeared in 1951, and Neoplan introduced fresh-air nozzles (*à la* commercial airliners) in 1960. Air suspension systems, first shown in Europe by several bus manufacturers at the 1957 IAA but long customary in American bus practice, gradually made their way into production. For trucks, their introduction was delayed by technical problems—until 1965 for Büssing and Berliet, and 1967 for Rheinstahl-Henschel and MAN. Retarder brakes and disc brakes appeared in buses before they were installed in trucks. Bus makers had not only declared their independence from truck design ethics but had also assumed the technical lead.

Beginning in the early 1960s, several commercial vehicle manufacturers ceased production—Borgward and German Ford in 1961, DKW in 1962/1975 (1962 in Germany, 1975 in Spain), Opel in 1975, and Bedford/GM in 1986. Other firms underwent takeovers, mergers, and alliances. The most important of these were as follows: Pacific Car & Foundry/Paccar took over Kenworth in 1945, Peterbilt in 1958, Foden in the 1980s, and finally DAF in 1997. Hanomag became part of Rheinstahl in 1952, to be joined by Henschel in 1964. Hanomag-Henschel in turn became part of Daimler-Benz AG in 1970, after Daimler-Benz had already taken over Krupp in 1968. Gutehoffnungshütte/MAN took over Büssing in 1969/1971, Steyr in 1989, the Polish firm Star in 1999, and ERF of Britain in 2000. Magirus, Saviem, DAF, and Volvo formed their own alliance in 1971.

In 1975, Fiat's IVECO absorbed Lancia, OM, Unic, Magirus, and later Ford/UK. Renault added Berliet and Saviem to its collection in 1976, forming Renault Vehicules Industriels (RVI), which took over Mack Trucks in 1983. Renault sold its RVI operations to AB Volvo in 2000. In 1981, Mercedes and Volvo took over Freightliner and White, respectively. In 1987, Leyland Trucks and Freight Rover went to DAF, while control of Drögmöller passed to Volvo in 1994. In 1995, Kässbohrer, now called Evobus, came under Mercedes control, while the bus divisions of RVI and IVECO joined forces in 1999 to create Irisbus. In 2000, Volkswagen acquired 34 percent of Scania.

Of the 127 German firms that have built commercial vehicles at one time or another during the past 104 years, only a handful remain. Medium and heavy trucks are built by MAN, DaimlerChrysler, and Volkswagen. Delivery trucks are provided by Ford of Germany, DaimlerChrysler, Volkswagen, and Opel (based on Renault). Added to these are several smaller firms that produce special bodies and equipment. As is the situation in the bus industry, the Big Two—DaimlerChrysler with Evobus, MAN with Neoplan/Gottlob Auwärter—enjoy worldwide operations, while five small suppliers fill various market niches.

In 1960, government regulations prohibited towing passenger-carrying trailers behind buses. Combined with the phase-out of unpopular streetcar systems by several municipalities, German communities were forced to find other solutions to fulfill metropolitan transit needs. In addition to commuter rail and subway systems, articulated buses found their way into large cities. Similar to streetcars, these had a capacity of approximately 165 passengers. Initially, the usual layout consisted of an underfloor engine in the front unit and a driven center axle (B axle). Beginning in 1977, these were replaced by "pusher" articulated buses with an engine mounted at the back of the rear unit, driving the single axle (C axle) of that unit. Electronically controlled anti-jackknife devices for the articulation joint, first developed in Germany by Fahrzeugwerkstätten Falkenried (Falkenried Vehicle Shops), eliminated the inherent instability of the pusher configuration (Figure 188).

Figure 188. FFG articulated pusher bus, 1976. An electronic anti-jackknife device developed by Fahrzeugwerkstätten Falkenried (Falkenried Vehicle Works) of Hamburg offers advantages for both passengers and bus manufacturers: lower floor, lower noise levels, and precise tracking through turns. Because the sections consist of one standard bus without a rear wall and another without a front, the concept lends itself to rational production on standard bus assembly lines. Mercedes-Benz licensed the concept (1977 Mercedes O 305 G).

To eliminate model proliferation and rationalize local public transit, German and other European bus manufacturers adopted guidelines set down in 1966–1967 by the VÖV, the Verband Öffentlicher Verkehrsbetriebe (Association of Public Transit Systems), for "standard transit buses." For the next two or three decades, these defined two standard configurations. First was the standard city transit bus (beginning in 1968) with double doors front and rear, low floor, and stepdown, standing room, and characteristic bulging windshields (Figure 188).

The other design direction resulted in the tour bus, emphasizing comfort, in a high-deck configuration with approximately 10–15 m^3 luggage capacity, on-board galley/bistro/office, driver's sleeping bunk, and a restroom/toilet on the lower deck. The high-deck design goes back to the mid-1930s Super Coach 719 by Yellow, continued by Kässbohrer in 1956–1958 to specifications laid down by Continental Trailways for its North American operations (Figure 189) and introduced to European operations by Neoplan in 1971. Neoplan, which also produces buses in the United States, pioneered many aspects of bus technology, layout, and design (Figure 190): front-drive airport ramp buses in 1960, double-decker tour buses in 1967, articulated double-decker buses in 1975, high-deck buses with an underfloor cockpit in 1979, low-floor city transit buses with carbon fiber reinforced unit bodies in 1988/1991, low-floor articulated trolley buses with wheel hub motors in 1992, and many other innovations.

Figure 189. Kässbohrer Setra high-deck bus, 1956–1957. Under competitive pressure from airlines, American cross-country bus operator Continental Trailways commissioned the Kässbohrer vehicle works to build approximately 200 high-deck buses. The Trailways production contract ended in 1961. The high-deck design— with underfloor space for luggage, parcels, and freight—was a configuration unfamiliar to German operators at the time. It marked the first use of a German bus in American transcontinental service. Six-cylinder turbocharged MAN engine, 8276 cc, 155 hp (114 kW).

Figure 190. Neoplan Starliner, 1997. Following the examples set by its 1961 "Hamburg" and 1971 "Cityliner" designs, the 1997 Starliner again represented a definitive new tour bus design. A glass canopy instead of a windshield, visual emphasis on a triangular B-pillar with integral roll bar, "invisible" window frames, side windows wrapped into the roof, and covered rear wheels give this technologically advanced bus a futuristic look. V8 diesel engine (Mercedes), 14,618 cc, 381 hp (280 kW).

In the 1990s, bus manufacturers departed from the VÖV guidelines and returned to their own company-specific designs. At the same time, combination buses were developed, allowing operators to earn revenue using the same vehicle for scheduled transit service as well as charters, tours, and point-to-point service. In the process, transit and overland buses grew to the maximum permissible length of 15 meters (49.2 ft). To operate in narrow city streets, these buses were fitted with a third axle and electronic steering system (Figure 191).

After pneumatic tires and diesel engines, electronics represent the third significant development in truck and bus technology. Anti-lock braking systems (ABS), offered by Mercedes and Wabco-Westinghouse in 1981 and by MAN and Bosch-Knorr in 1982 (Figure 192), were the first electronically controlled systems to find application in the commercial vehicle sector. These were followed by electronic controls for transmissions, traction control (anti-slip regulation, ASR), fuel injection pumps, and disc brakes. Work is underway on electronic steering (steer by wire). Trailer-mounted electronics, which can transmit information on parameters such as cargo bay temperature, tire pressure, load condition, and vehicle location to the trucking company dispatcher, are already being offered as options. Another system, known as electronic stability program (ESP), can brake individual wheels in the event the vehicle threatens to deviate from its course. These are already being offered on delivery vehicles and will appear on heavy trucks in a few years. With ABS, ASR, and EBS (electronic brake systems), ESP will give trucks and tractor-trailers a complete vehicle dynamics control system.

Figure 191. Neoplan Transliner steering system, 1997. Thanks to electronic steering systems, 15-meter (49.2-ft) buses are almost as easy to maneuver through city traffic as conventional 12-meter (39.4-ft) transit buses. For example, the drive axle can turn 5 degrees, and the trailing axle can turn up to 25 degrees. Safety is also enhanced because on-board electronics react faster than the human driver to stabilize a skidding bus.

Figure 192. ABS anti-lock brakes for commercial vehicles, 1981–1982. As on passenger cars, commercial vehicle ABS, introduced by Mercedes and Wabco-Westinghouse in 1981 and by MAN and Bosch-Knorr in 1982, must guarantee directional stability, steerability, and shortest possible braking distances, even on critical surfaces. The marked wheels of the tractor/trailer combination shown here are prevented from locking, which eliminates the danger of jackknifing. Without ABS, the vehicle would skid with locked wheels.

Under pressure from tighter exhaust emissions standards, applying to commercial vehicles as well as automobiles, diesel engine manufacturers were forced above all to improve their vehicle injection systems. Previous efforts to increase engine efficiency with traditional methods resulted in higher nitrous oxides and soot emissions, which, in the eyes of the European

Union Environmental Commission, contribute to summer smog and acidification of ground and water. Diesel soot is also a suspected carcinogen.

To meet American exhaust emissions standards that took effect in 1974, Cummins Engine Company reintroduced unit injectors (see Appendix, Diesel fuel injection), which had been introduced to the U.S. market by General Motors in 1936. With unit injectors, an additional cam lobe on the overhead camshaft of the engine provides the necessary pressure for the unit injector, a significantly higher pressure than is used in conventional diesel inline injection pumps. In association with other design features, engine efficiency increased while nitrous oxides were reduced. These "big cam" engines were used in several products, including the Ford Transcontinental, sold in Germany as of 1975 (Figure 193).

Figure 193. Ford Transcontinental, 1975. At the time of its introduction, the Transconti, developed for the European market and fitted with a Cummins engine, featured unit injectors, turbocharging, and abundant power in the lower rev ranges. Furthermore, it was built using lean production methods and outsourcing long before these became industry buzzwords: Fuller transmission, Rockwell rear axle, and Berliet cab, after designs by Lepoix. Six-cylinder Cummins turbo diesel, 13,843 cc, 320 hp (235 kW).

In Europe, national emissions standards had to be coordinated before the rather mild standards of 88/77/EEC could be applied. The situation became more serious with the introduction of Euro 1 standards in 1993, which reduced NO_x limits by approximately 40 percent. This could be achieved with late injection, lower combustion temperatures, and higher fuel consumption (i.e., with lower thermodynamic efficiency). In 1993, only Volvo was in a position to reduce both NO_x and fuel consumption, by using unit injectors and electronic controls—both items representing new technology for a European manufacturer.

Tighter exhaust emissions standards to meet Euro 2 (1996), Euro 3 (2000), Euro 4 (2005), and Euro 5 (2008), each taking effect in October of the given year, have led to new or improved injection systems (see Appendix, Diesel fuel injection). Eventually, however, to meet these standards, it will be necessary to take steps beyond the engine itself (e.g., diesel fuels with higher cetane numbers and lower sulfur content).

Maintenance Methods: From Village Blacksmith to On-Board Diagnostics

If the truth were to be told, on early automobiles, everything broke—sooner or later. Automotive engineering was a vast unexplored territory. The main sore spot was tires. Around 1900, tires lasted only about 4000 km (2,500 miles) and, because of horseshoe nails littering the roads, were forever being patched. Ignition systems and carburetors also were a never-ending source of aggravation. Faulty design, badly dimensioned parts, and unsuitable material put many motorcars in the ditch, when, for example, steering arms or tie rods broke or the brakes failed.

In extreme cases, auto manufacturers sent traveling mechanics on the road—men whose task it was to create and install hand-made parts on the spot. Often, though, the skills of the local blacksmith were sufficient to return a broken-down motorcar to the road. Blacksmith shops, similar to sewing machine or bicycle repair shops, formed the basis of today's official factory repair shops and dealer and independent garages (Figure 194).

Because automobiles represented complex technology and were difficult to operate, most owners hired a chauffeur. The first chauffeurs were mechanics and assemblers, trained by factories, and "so to speak delivered to the customer along with the car" (Julius Küster, *Chauffeur-Schule*, Berlin, 1910). Around 1900, professional drivers came on the scene—men who had obtained their technical training and their knowledge of etiquette toward nobility in dedicated chauffeur schools.

The actual structure of the automotive world, following American practice, did not crystallize until the 1920s. Bosch created the first specialized automotive repair shops in 1921 (dealing with electrical repairs). In 1922, Olex opened the first German street gas station, in Berlin. Soon, the dispensing of gas from hand-pumped barrels in back-alley shops became a thing of the past (Figure 195).

Figure 194. Automobile repair shop, circa 1905. The repair shop of what would one day become Raffay & Co. of Hamburg, a major multi-brand (Volkswagen, Audi, Porsche, and SEAT) dealership with 15 locations. In 1905, this garage serviced the French marques Richard Brasier, Renault Frères, Clement-Bayard, and de Dion-Bouton. Apparently, one of these is being repaired here. The machinery is powered by lineshafts and belts; a crane and chainfall take the place of a pit.

Figure 195. General store with curbside gasoline pump, circa 1924. A housewares store for glass, porcelain, and miscellaneous goods, undergoing a transition to selling more "technical" items such as cream separators and butter churns, automotive spare parts, fuel, and lubricants. The gas pump ("Iron Maiden") shown here is a second-generation curbside pump, seen in Germany beginning in the early 1920s.

Following the example of Ford in the United States, Opel covered the German Reich with a network of dealers and factory authorized repair shops, introduced fixed prices for repair work, offered warranties on factory-original spare parts, and in 1931 opened a customer service school in Rüsselsheim (Figure 196).

Figure 196. Opel service school, 1931. An overabundance of marques and types, built in limited numbers, overwhelmed blacksmiths and master mechanics who had entered the auto repair business—to say nothing of the less reputable practitioners. In the 1920s, auto makers themselves recognized that the situation endangered the entire industry. Following American practice, several German auto plants, led by Opel, built up their own service organizations. Opel founded Germany's first service school in 1931.

The gradually rising middle class could afford a bicycle, perhaps a motorcycle, or at best a middle-class car, but certainly not a chauffeur. Therefore, auto manufacturers issued handbooks that outlined minor repair procedures. These books usually were divided into sections for operations (for the customer), maintenance (for the customer and the repair shop), and repairs and adjustments (for the repair shop). To this day, auto makers continue to bless us with an owner's (operations) manual. However, since the 1950s, maintenance and repairs have been relegated to workshop manuals.

The 1930s saw rapid growth in the number of factory-authorized repair shops. Specializing in a single marque had advantages for all parties. First, auto makers gained presence and sales throughout the country. Likewise, the shops earned additional revenue through sales of new cars and accessories, and received financial aid, and technical and sales training from the factory. Furthermore, customers received expert repairs. Increasingly, repair shops replaced their street gasoline pumps with full-fledged gasoline stations and additionally operated repair garages and driving schools. Full-service, American style, had been born.

Ultimately, these businesses shifted their concentration from repair to preventive maintenance. To this end, in the 1930s and again in the 1950s, auto makers introduced customer service "coupon books" with recommendations for lubrication, oil change, and tune-up and adjustment intervals. Because of numerous lube points and maintenance-intensive technology, customers had to bring their cars to the shop for frequent care and service but at least could theoretically count on trouble-free motoring and higher resale value of the vehicle. Even in the 1930s, the word "repair" held negative connotations; the industry preferred to call it "service."

In Europe, after World War II, dilapidated or destroyed repair shops had to be rebuilt and modernized, and repair processes rationalized.

One such rationalization method was the introduction of exchange parts. The trade regarded this with mixed emotions. On one hand, customers could have their cars restored to operating condition in short order, even after major damage. On the other hand, repair shops lost a lucrative part of their business. Unlike the 1930s, when exchange parts were limited to small units such as fuel pumps, starters, or generators, they now included more expensive and labor-intensive components such as engines, transmissions, and rear axles.

Beginning in the mid-1950s, major repair facilities reacted to the increasing scope of repair and maintenance work by separating this from regular shop activities, creating "assembly line" inspection processes. Here, several mechanics worked simultaneously on the same car, each with rigidly defined tasks. For example, Volkswagen reduced the time and labor required for normal service from one mechanic working two hours to three mechanics working for 30 minutes, four mechanics working for 23 minutes, or a crew of five working for 17 minutes (Figure 197).

In normal shop operations, individual test and repair operations such as engine power checks, electrical system, wheel alignment, brakes, body, and final inspection were all given their own dedicated shop space, equipped with the necessary tools and test and adjustment equipment, supplemented by pits, hoists, work platforms, and chassis dynamometers.

A scheduler handled work assignments, job coordination, procedures, and deadline monitoring. The scheduler communicated with various work stations, the master mechanic, and service writers via pneumatic mail system, public address system, and telephone (Figure 198).

Instead of the master mechanic, customer contact was now the domain of the service writer, who established the scope of the work to be done and the job scheduling. The master mechanic's formerly dominant position was now under attack from two directions: from within, by the

Figure 197. Routine service, assembly-line style, circa 1955. Several large Volkswagen dealers, such as Berlin's Eduard Winter seen here, introduced assembly-line concepts to routine inspection and maintenance. All vehicles, regardless of type—Beetle, Karmann-Ghia, or Transporter—moved along the line at a steady rate, passing through six work stations, staffed by 14 mechanics with rigidly defined tasks.

Figure 198. Repair shop with scheduler's office, circa 1983. The elevated scheduling office gives the dispatcher an overview of the entire shop. He is connected to individual service bays and the service writers in the main building via an intercom system, telephone, and pneumatic mail tubes. Truck and passenger car departments are separate, as shown here at a Mercedes dealership.

scheduler, and from without, by the customer service writer. By his demeanor, appearance, and specialized knowledge, the service writer had an even greater influence on customer perception of shop quality than did waiting times or service bills.

Customers themselves were not welcome in the shop. Their verbal orders were transmitted to the mechanics in the form of written work orders, via service writer and shop foreman or master mechanic. Imprecise work orders virtually guaranteed misunderstandings, high costs, and failure to meet job deadlines. This, and the anonymity of large shops, prompted more than a few customers to seek independent, specialty, or back-alley shops, or to try something new for German consumers—the do-it-yourself school of auto maintenance.

Since the 1970s, two external developments have left their marks on the auto repair trade: long-life components and lubricants, and automotive electronics. Even after 30 years, their effects on the structure of the trade are not yet fully known.

With the aid of improved components and better materials and lubricants, the amount of scheduled maintenance has decreased, and maintenance intervals have increased—from once every 1500 km (932 miles) in the 1950s to 10 or 20 times that interval or more. Since the late 1990s, scheduled maintenance intervals for trucks have been set at 100,000-km (62,140-mile) intervals. The result has been that despite more vehicles on the road, repair shops are not working to full capacity.

Because of electronic components and test systems, in the car and in the shop, service and repair costs will continue to decrease. The impetus toward electronic automotive test equipment originated in the United States. In 1962, Mobil Oil opened the first diagnostic center. Because Europe was a much less affluent market, such equipment was deemed too costly an investment, and smaller and less capable engine testers were used instead.

Microprocessor-controlled diagnostic devices have been in service since around 1980. These no longer display information on oscilloscopes but rather indicate observed and target values on computer monitors, digital readouts, or LEDs. Diagnostic programs help localize and eliminate faults in the electrical system of the vehicle (Figure 199).

Faults and deviations from optimum conditions can be localized by means of on-board diagnostics (OBD; see Appendix) and reported to the driver by means of warning lights or displays. OBD II, required in the United States since 1994, monitors components that affect exhaust gas composition (e.g., catalytic converter, oxygen sensors, fuel injection, ignition, fuel system ventilation, and exhaust gas recirculation). Using a fault code buffer and diagnostic equipment, a service technician can go directly to the source of the problem. The cost of time-intensive troubleshooting is no longer passed on to the customer.

If OBD technology were required in Europe, Germany could do away with its annual exhaust emissions test (Abgasuntersuchung, AU). Future OBD technology could be expanded to

Figure 199. Diagnostic electronics, circa 1985. After initial negative experiences with diagnostic computers, fraught with too many technical and personnel difficulties (Volkswagen, 1971–1978), repair shops returned to a more highly developed modular system around 1980. Microprocessor-controlled diagnostic devices compared individual test parameters with specifications, pinpointing the sources of problems.

encompass other electronic systems within the car. It is conceivable that fault codes might be transmitted automatically to a monitoring organization (in Germany, the TÜV, Dekra, or other institutions), to an authorized shop, or to the manufacturing plant. OBD technology might eliminate the need for routine inspections and the biennial vehicle inspection required by Germany's §29 StVZO (Section 29 of the Motor Vehicle Code).

Alternative Fuels, Engines, and Drive Systems

Historically, the internal combustion engine powered by gasoline or light distillates was by no means the first means of propelling a road vehicle. However, in contrast to its predecessors, steam and electricity, it was easier to operate and offered greater range.

The advantages of the gasoline engine, however, have been challenged repeatedly. Examples include decades of work by Rudolf Diesel and MAN with "gas oil," which ultimately led to the successful diesel engine. The problem posed by the coal dust engine, proposed by Diesel in 1893 and 1899, remains unsolved. Further research into dust-powered engines by Rudolf Pawlikowski (Rupa engine), and by the firms F. Schichau AG and IG Farbenindustrie AG, employing coal dust or other organic wastes such as sawdust, rice husks, and turf, ended around 1940. Researchers found it impossible to keep hard residue or ash particles out of the lubricating oil.

Under the exigencies of war or economic hard times, town gas (illuminating gas), bottled gas (propane-butane mixture), and producer gas systems using wood or low-temperature coke gained a certain importance. To assure production and supply of liquid fuels during war, as early as September 1939 the German government began rationing gasoline and diesel fuel used for private purposes, and simultaneously ordered conversion of trucks (Figures 135 and 200) and buses (cf., Figure 118) to operate on liquefied gas or producer gas generators.

Figure 200. Vomag 5 Cz/6 EH with Imbert gas generator, 1934. Instead of being burnt, wooden logs are gasified in the cylindrical gas producer (generator). The gas is fed to a precipitating tank (below the bumper), where it is cleaned and allowed to expand. The gas is cooled by vaporizing water in the gas cooler, located immediately above. Next it flows through a filter made of pulverized cork (not visible), through an air/gas mixer or carburetor to the combustion chambers of the engine. Six-cylinder engine, 9966 cc, 80 hp (59 kW) on gasoline, roughly 40 percent less in producer gas operation.

While conversion of transit buses to liquefied, high-, or low-pressure gas was relatively straight-forward, outside city limits only producer gas generators were suited to the task. Commonly known as "wood gas" generators or "Imberts," they had been used on occasion in France and Germany since the 1920s but during World War II experienced unprecedented growth. Along with a small number of generators by other manufacturers, George Imbert's gas generator was the only functioning alternative energy source in areas lacking conventional fuel or whose supply infrastructure had been destroyed. Through 1948, Imbert made approximately 532,000 of these units for road and rail vehicles and watercraft.

In the 1950s, the automobile industry investigated more advanced alternative drive systems for road vehicles. Significantly, neither steam nor electric propulsion received any further attention. Enough data had already been accumulated on these systems, and no further research was needed. Instead, beginning around 1950, several European (e.g., Rover, Fiat, Renault, and Gregoire/ SOCEMA) and American firms (e.g., Chrysler) experimented with gas turbine-driven passenger cars (see Appendix). Several experimental and record-setting cars were built.

Gas turbine drive for trucks seemed to hold greater promise. In the 1960s, Kenworth and Ford in the United States, and Leyland, Magirus, MAN, and others in Europe had experimental turbine-powered vehicles in operation. Because of excessively high manufacturing costs and fuel consumption, as well as real-world operating conditions, gas turbines never moved past the prototype stage. The only gas turbine-powered land vehicle to go into production was the American M1 "Abrams" main battle tank, which ceased production in the late 1990s.

Another alternative to the reciprocating engine is the rotary piston engine developed by Felix Wankel (1902–1988, see Appendix). The world's first production rotary piston engine, the NSU-Wankel Spider, came on the market in 1964, in cooperation with NSU (Figure 170). In Japan, Toyo Kogyo (Mazda) had obtained Wankel licenses even earlier (1961) and in 1967 introduced its Cosmo sports car, powered by a twin-rotor Wankel engine. That same year, NSU began production of the Ro 80 (Figure 163), the first production sedan powered by a rotary piston engine. Citroën became the third firm to take up manufacturing Wankel-powered cars, with its Birotor of 1974.

Until around May 1974, approximately 30 licensees throughout the world investigated Wankel rotary engines. At that point, the field saw a sudden turnaround: General Motors and Daimler-Benz both closed major Wankel projects. Peugeot, parent of Citroën, halted Birotor production after only one year. Volkswagen, as parent of Audi/NSU, took the Ro 80 out of its program in 1977. Almost all licensees, with the exception of Mazda, terminated work on the Wankel engine. In 1982, the Norton-Villiers-Triumph Group began production of a Norton motorcycle, powered by a rotary piston engine.

What went wrong? By the time of the 1973–1974 energy crisis and under pressure from exhaust emissions and fuel economy restrictions, American car makers finally were forced to consider replacing their large models with more economical designs. This required major capital investment. The added expense of introducing a new engine concept would have stretched the financial resources of individual auto firms beyond acceptable limits, or at least had a negative impact on stock dividends. Thus, the United States, or more accurately General Motors, the market leader, dropped the rotary engine, and the remainder of the world followed suit. The Wankel rotary engine did not falter because of technical difficulties but rather because of the unfavorable geopolitical and corporate situation of the time.

Undaunted, Mazda continued to build Wankel-powered cars. By the end of 1978, the Cosmo sports car had been followed by 12 different models, all powered by rotaries, including sedans, pickups, and buses. In that year, Mazda introduced its RX-7 (Figure 201), which would become one of the most successful sports cars of all time and which remains in production today. Reporting Mazda's most significant racing victory in its June 27, 1991 issue, the Swiss weekly *Automobil Revue* commented that "With its 1991 Le Mans victory, this engine concept has achieved its most notable success to date." More importantly, a relatively small firm with an even smaller racing budget had beaten well-endowed competition powered by reciprocating engines, including Mercedes, Porsche, and Jaguar, and their decades of racing experience—a unique event in the history of motorsports as well as the history of technology.

Figure 201. Mazda RX-7, 1980. In production since 1978 and continuously improved to the present day, the RX-7 sports car, powered by a twin-rotor Wankel engine, now develops as much as 280 hp. The RX-7 is no longer offered in the United States or Europe but was kept alive in 1999 in the form of a concept car, the rotary-powered RX-Evolv sport sedan. Twin-rotor Wankel engine, chamber volume 2×573 cc, 105 hp (77 kW).

Mazda sees a bright future for hydrogen-fueled rotary engines. In contrast to conventional reciprocating engines, stable continuous combustion is achieved by means of a simple modification.

Besides Mazda, other firms have spent decades working with hydrogen fuel for road vehicles, admittedly with piston engines. The most tempting features of hydrogen are independence from fossil fuel sources and its environmental friendliness. When burned, hydrogen and oxygen combine to form water, with almost no pollutants. However, hydrogen is extremely volatile and is most reasonably transported in its liquid state at a temperature of –250°C (–418°F). Presumably, robots would handle the refueling process. Because it contains only one-third as much energy as gasoline, hydrogen cars would require a voluminous fuel tank, which will negatively impact the space and weight situation.

To date, there has been no satisfactory answer to the definitive question of how hydrogen, which does not appear freely in nature, can be produced in an environmentally neutral and economical manner. At present, it is produced from hydrocarbon compounds—petroleum and natural gas—or by electrolysis of water, which consumes a great deal of electricity. Either way, hydrogen production is environmentally harmful because carbon contained in fossil fuels is released to the atmosphere as carbon dioxide. It will be many years before solar thermal powerplants produce so much electricity that hydrogen can be regarded as an economical byproduct.

Today, alternative fuel research is driven not by fear of dwindling oil reserves, loss of self-sufficiency, or colonial, economic, or political considerations, as had been the case until around 1990 (e.g., alcohol-powered cars in Brazil), but rather by environmental concerns. Emissions from the entire energy chain must be taken into consideration—from obtaining raw material through processing, transportation, storage, and refueling. Local improvements in air quality cannot be traded off against higher emissions at another location. If the complete chain is considered, an entire range of alternative fuels and propulsion systems does not necessarily have the ecological advantages that its proponents would have us believe.

According to a study by Prognos, a research and consulting firm based in Basel, Switzerland, in 1992 a midrange piston-engined car, fueled by liquid hydrogen derived from natural gas, had twice the overall energy consumption as a similar, conventional, gasoline-fueled vehicle. If the hydrogen fuel were obtained by electrolysis using electricity from the national grid, its overall energy consumption would be quadruple that of the gasoline-powered car. In Germany, for example, in 1991 the national grid obtained 32 percent of its power from nuclear energy, 31 from lignite (soft coal), 26 percent from anthracite, 5 percent from natural gas, 3 percent from hydropower, 2 percent from heating oil, and 1 percent from other sources.

Even if the relative percentages have shifted slightly in favor of renewable resources such as wind, solar, or hydro- or photovoltaic power, hydrogen-powered vehicles will remain ecological and economic absurdities unless and until an economical, non-fossil-fuel source of electricity becomes available.

More rapid introduction of hydrogen fuels is conceivable if crude oil prices rise drastically, if climatological changes impact even the temperate zones, or if zero-emissions or low-emissions vehicles are mandated by law. The last option has already taken place.

In 1990, the California Air Resources Board (CARB) ordained that as of 1998, 2 percent of all vehicles sold by any manufacturer must meet so-called ZEV (Zero Emissions Vehicle) standards (i.e., vehicles with absolutely no exhaust emissions whatsoever). The proportion of ZEVs was to rise to 5 percent in 2001 and to 10 percent as of 2003. The rule was to apply to manufacturers selling more than 35,000 cars per year in California.

American car makers General Motors, Ford, and Chrysler, as well as those Japanese firms with U.S. plants (i.e., Toyota, Honda, Mazda, and Nissan) reacted with a two-pronged strategy. On one hand, they were up in arms against targets and timetables that they regarded as totally unrealistic, with the result that the ZEV rules were relaxed in 1996 into self-enforced industry goals. On the other hand, they and other auto makers investigated alternative fuels and drive systems: electric, hybrid, and fuel cell power; liquefied petroleum gas; natural gas; fuel from biomass; and reformulated gasoline and diesel fuels.

In the case of electric cars, the vehicle is powered by one or more electric motors. The required energy is carried on board in the form of batteries (battery-electric power) or obtained directly from sunlight by solar cells (solar power). Electric cars do not produce any emissions while in operation and are quieter than cars powered by internal combustion engines. Electric motors are maintenance-free and can exceed the service life of combustion engines by a considerable amount.

General Motors produced, among other electric cars, its "Impact" experimental vehicle, later produced as the Saturn EV1. Chrysler built an electric van, Ford converted a European Escort delivery van to electric drive, and Honda and Fiat built subcompacts, to name only a few of the electric-car projects. Renault and PSA held the technological lead. They had already begun to work with electrics long before the introduction of the U.S. Clean Air Act. By the mid-1990s, they were able to offer complete systems (Figure 202).

Figure 202. PSA Tulip System, 1995. PSA intended its Tulip, an electric car measuring 220 cm (87 in.) overall length, not only as an automobile but as an addition to existing personal and public transportation systems. Following the example of Amsterdam's Witkar system of the mid-1970s, these two-seat rental cars could be used by anyone and then parked in specific recharging stations. Community trials were planned, but at present part of the Tulip fleet resides in museums.

By 1999, it was obvious that the auto-buying public would not accept electric cars. Of the 2,465 units sold on the U.S. market between 1996 and mid-1999, most were purchased by public utilities that were required by law to use low-emissions vehicles. In 1999, Honda withdrew from the electric vehicle market, after investing millions and selling only 300 units. The remaining six U.S. manufacturers—General Motors, Ford, DaimlerChrysler, Nissan, Toyota, and Mazda—soldier onward, if only to keep up appearances. Even in the early 1990s, it was apparent that no demand for electric vehicles existed, capital invested in electrics would not be recovered, and that any success of electric vehicles would be deferred pending a hoped-for breakthrough in battery technology.

To date, this has not been realized. As early as 1898, *Der Motorwagen* (Vol. 9, p. 90) expressed the hope that "in the coming century, electricity will be the driving force for elegant fiacres [a type of carriage] and for luxury town cars, especially if progress continues to be made in the manufacture of accumulators [batteries]." These advances did not materialize. Significant progress has yet to be achieved. Then as now, the fate of electric cars hinges on batteries.

Batteries are not only heavy and bulky, but are also expensive and can store only limited amounts of energy. This impacts the performance and range of electric vehicles. Recharging a battery takes hours, its life expectancy is at best 100,000 km (60,000 miles), and recycling costs are high.

For these reasons, pure battery power is of interest only for special situations, regardless of whether the application is a passenger car, truck, or bus. At least once every decade, it seems, the postal authorities of various countries attempt to operate electric vehicles in a cost-effective manner, only to fail. State-supported large-scale experiments suffer the same fate (e.g., on the island of Rügen, 1996), as do electric bus projects fostered by several communities. Pure solar operation (Figure 203) cannot be taken seriously for automotive applications because the efficiency of solar cells remains inadequate.

If it ever becomes possible to obtain electricity in an economical manner from non-fossil energy sources and to somehow optimize storage batteries, electric propulsion may someday replace the combustion engine. Until then, electrical power may be produced by an on-board generator. Such hybrid vehicles usually employ a combustion engine (e.g., gasoline, diesel, or liquefied petroleum gas) coupled to a generator. Geared transmission and clutch may be eliminated, and, in the case of wheel hub motors or motors installed near the wheels, driveshafts also become superfluous. The number of required batteries is reduced. In town, such vehicles operate under electric power. On highways or in the event the batteries are not charged, the combustion engine takes over.

A basic drawback is greater vehicle weight due to two separate drive systems—that is, an unfavorable ratio of curb weight to payload. However, recent developments have shown progress in rectifying this situation.

Figure 203. Solar cars. For approximately 25 years, efforts have been underway to employ photovoltaic cells—used since 1954 in space exploration, the photo industry, and architectural applications—as energy sources for road vehicles. Despite intense sunlight, 4 m² of collector area on the Mercedes/Alpha Real of 1985, for example, supplied only approximately one-third of the energy required to move the vehicle at 50 km/h (30 mph). The remainder had to be drawn from on-board batteries. Solar energy is not yet suitable as the sole power source for road vehicles. Two DC motors, 2×900 W, silver-zinc battery, 48V, 90 Ah.

Aside from submarines and diesel-electric locomotives, which have long used such mixed drive systems, hybrid land vehicles have an equally long tradition of their own. In 1901, Lohner of Vienna offered cars with "Mixte" drive, in which a generator, coupled to a gasoline engine, supplied electric power to wheel hub motors. By 1906, Lohner had managed to sell only a handful of these hybrid vehicles, which were designed by a young Ferdinand Porsche. Most of these were passenger cars, with a few trucks.

After Porsche was appointed chief engineer of the Austrian Daimler-Motoren-Gesellschaft in 1906, he had this firm build gasoline-electric vehicles as well, mainly fire engines with wheel hub motors (Figure 204) and, in World War I, military vehicles for road and rail service.

In the early days of the automobile, "mixed" drives were not developed to protect the environment but rather to eliminate clutch wear, difficult shifting, driveline shock loads, and broken gears. Because they lacked a direct mechanical drive, gasoline-electrics could start smoothly,

Figure 204. Austro-Daimler fire engine, 1910–1913. A gasoline-electric fire engine, with an electric generator mounted ahead of the radiator, feeding power to front wheel hub motors. The electric motors consist of radial field coils arranged around axle stubs. The hubs act as armatures. A rear-wheel drive configuration was also possible. Four-cylinder engine, 6963 cc, 60 hp (44 kW), 36 kW generator.

a desirable feature for community-owned vehicles making frequent stops and starts. Around 1930, Faun offered waste disposal trucks with gasoline engines and electric hub motors at the rear axle.

More recently, bus design took the lead in mixed drive development. In May 1983, the city of Essen placed into service Mercedes buses that used overhead trolleys while operating in the city center, and switched to diesel engine operation when outside the downtown area. In December 1993, Neoplan began deliveries of hybrid buses that employed a diesel motor/generator unit and control electronics to power hub motors in the rear wheels (Figure 205). Later, these buses were fitted with flywheel storage systems or batteries to enable them to operate over several kilometers with the diesel engine shut down.

Passenger car makers are also actively engaged in hybrid engine research. After many prototypes by various manufacturers, Toyota became the first car maker since Lohner in 1906 to introduce a production hybrid automobile—the Prius of 1997. The Toyota system combines the two most familiar drive systems: power produced by a gasoline engine is split, part going through an epicyclic transmission to the front drive wheels, and the remainder driving a generator which in turn feeds an electric motor. Under load, the electric motor provides added power. The combustion engine and electric motor work simultaneously to propel the vehicle

Figure 205. Neoplan Metroliner N 8012 DE hybrid bus, 1993. A diesel engine mounted transversely at the rear, directly coupled to a generator, is electronically controlled to feed two electric motors, one in each rear wheel (MM Magnet-Motor Starnberg system). This eliminates the need for a transmission and halfshafts, and gives wide latitude in locating the diesel engine. The unit body is a carbon fiber and fiberglass composite structure. Six-cylinder Deutz diesel, 4790 cc, 169 hp (124 kW).

(parallel hybrid operation). In low-demand situations, the combustion engine shuts down, and the vehicle is driven only by batteries and the electric motor (serial hybrid operation).

While Honda's Insight is already in production, Fiat (Multipla) and Citroën (Berlingo Dynavolt and Xsara Dynactive) are at present engaged in the final phases of prototype development and production startup. Audi, after a pre-production series of approximately 50 of its Duo model (based on the A4), canceled its hybrid project in the second half of 1998 because it regarded the sales potential as insufficient.

The aforementioned drive systems occupy a definite market niche and therefore will certainly enjoy at least limited sales. However, the auto industry hopes that future road vehicles may be powered by fuel cells, which promise to make true zero emissions vehicles a reality. Fuel cells (see Appendix) employ electrochemical oxidation ("cold combustion") of a fuel with atmospheric oxygen to convert chemical energy directly into electrical energy. The mechanical detour through a combustion engine is eliminated. Fuel cell efficiency is approximately 50 percent, compared to approximately 37–45 percent for diesels and 32–40 percent for gasoline engines. Hydrogen serves as fuel.

Systematic research into galvanic fuel elements began with theoretical work by Nobel laureate Wilhelm Ostwald, who as early as 1894 proposed replacing inefficient heat engines with fuel cells. Until the end of the 1930s, fuel cell research was conducted mainly in Germany.

After the war, research and development into stationary generating systems shifted to other nations, mainly the United States.

The first land vehicle to be powered by fuel cells was quite possibly a 1959 Allis-Chalmers tractor, with General Electric cells. This was followed in the early 1960s by the U.S. Army M37 trucks with Monsanto fuel cells, the General Motors 1966 "Electrovan" delivery vehicle (Figure 206) with Union Carbide equipment, and, in 1975 in response to the first oil crisis of 1973–1974, the tiny DAF 44 with a Shell-Lucas powerplant. After that, development stagnated, at least with regard to vehicular drive systems. Advantages, such as emissions-free operation with hydrogen fuel, as well as low noise levels, did not outweigh disadvantages such as high operating cost and weight.

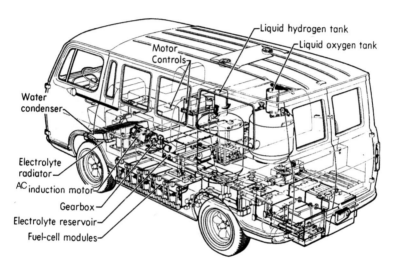

Figure 206. General Motors Electrovan, 1966. A rolling testbed based on a production GMC Handi-Bus, fitted with oxygen and hydrogen tanks, fuel cells, water condenser, and AC motor. Because the motor spun to 13,000 rpm, a stepdown transmission was required. Drive was through the rear wheels. The vehicle weighed 3220 kg (7099 lb), compared to a normal production van weight of 1449 kg (3194 lb). 115 hp (85 kW); production gasoline engine 140 hp (109 kW).

Fuel cells as a vehicular power source were awakened from their slumber by zero emissions standards promulgated in 1990. Several firms made good use of technology developed for military or space applications. One of these was Mercedes-Benz AG, whose aerospace subsidiary Dornier contributed its expertise to Mercedes' first NECAR (New Electric Car) project (1994). This consisted of a small MB 100 BZ delivery van. Similar to the General Motors Electrovan 18 years earlier (Figure 206), its cargo space was packed with the hardware required by this new propulsion system. Around the same time, Ballard Power Systems, a Canadian firm, showed a 10-meter (32.8-ft) bus, also employing hydrogen as fuel.

Because it is assumed that a widespread hydrogen supply infrastructure will not be available anytime in the foreseeable future, the auto industry is leaning toward an intermediate solution that uses methanol, which can be delivered in much the same way as gasoline. However, methanol will not operate without harmful emissions because it is an alcohol derived from methane, which is itself a constituent of natural gas and therefore a hydrocarbon-based non-renewable resource. The whole idea, after all, was to replace non-renewable fossil fuels with hydrogen fuel cells.

However, the latest prototypes by DaimlerChrysler as well as experimental vehicles by General Motors, Ford (USA), Toyota, Renault, and Volkswagen are fitted with on-board catalytic reformers that derive hydrogen from methanol. It is hoped that by 2005, costs, dimensions, and weight of hydrogen propulsion systems can be reduced to the point where some thought can be given to mass producing such a vehicle. Hydrogen-powered cars employing methanol reformers will not meet zero emissions standards but presumably will produce lower emissions than gasoline or diesel engines.

Remember that the driving force behind any possible future automotive propulsion system did not originate with any individuals, companies, society in general, nor the auto industry, but rather by the California state government through its regulatory agencies. Benefits for stationary applications (e.g., power generation and household applications) are also desired. Indeed, in view of the ultimate goals of environmental protection, these are inevitable if the desired emissions and consumption reductions are to be achieved. Moreover, direct hydrogen operation without the mediation of methanol would reduce dependence on natural gas and crude oil.

A similar effect would be produced by Stirling engines (see Appendix). Invented in 1816 by Robert Stirling, a Scottish cleric, and made practical by John Ericsson of Sweden from around 1840, the Stirling hot-air engine employs external combustion (unlike the conventional reciprocating or rotary piston engine). This makes it possible to design the Stirling as a multi-fuel engine, fired by anything combustible right down to coal, wood, or peat. Indeed, the boilers of stationary hot-air engines, which began to replace atmospheric gas engines from around 1874–1875 in shops and factories, were fired by these fuels. They provided more power and could operate independently of the gas supply. However, after the Gasmotoren-Fabrik Deutz began producing "Otto's new engine" (i.e., the four-stroke engine based on the work of Nikolaus August Otto) in 1876–1877, the less economical Stirling engine sank into obscurity.

In the late 1930s, Philips of Eindhoven, Holland, again took up the cause of hot-air engines. Since that time, the firm has made stationary power sources for special applications including pumps, military communications, space travel, and solar energy. Low pollutant and noise emissions, good efficiency, and vibration-free operation hold the promise of automotive applications. Between 1970 and 1980, Ford, General Motors, MAN, and MWM tried Philips hot-air engines in buses, passenger cars, trucks, and even marine vessels. The Swedish firm United Stirling AB installed a V4 Stirling engine in a Ford Pinto. To date, however, expense, complexity, and long warm-up times have posed an obstacle to more widespread application of Stirling engines.

Development of new propulsion systems employing rotary piston engines, gas turbines, battery power, and hybrid and hot-air engines have consumed millions of dollars of investment funds and will continue to do so. However, during the past 140 years (Lenoir, 1860), a wealth of experience has been accumulated with conventional combustion engines. Therefore, it makes sense to improve the combustion engine or convert it to more suitable fuels to reduce emissions and fuel consumption.

Alternative fuels include liquefied petroleum gas (LPG), compressed natural gas, (CNG), liquefied natural gas (LNG), methanol, and biofuels.

LPG, consisting mainly of butane and propane, is produced from crude oil and thus is not a genuine alternative fuel. Its NO_x and CO emissions are comparable to those of natural gas. Approximately 4 million LPG vehicles are in operation worldwide (Shell, 1992)—a minuscule portion compared to a global passenger car and commercial vehicle fleet numbering approximately 737 million. LPG buses and cars are primarily employed in the United States, Holland, Italy, Australia, Japan, and Canada.

Natural gas is liberated in huge amounts during crude oil production and is usually burned off. However, with appropriately designed injection, ignition, and fuel supply systems, it is perfectly suited for use in combustion engines. It produces significantly lower combustion byproducts than gasoline but requires catalytic converters.

Disadvantages include limited range, storage in high-pressure vessels in conventional CNG operation, and the danger of leakage. Methane contained in natural gas is approximately 35 times as harmful to the environment as carbon dioxide. LNG operation requires less space for the fuel tank, but the tank must be cooled to $-163°C$ ($-261°F$). Approximately 1 million natural gas vehicles are in operation (Shell, 1992), mainly in Russia, Argentina, and Italy. Germany has approximately 6,000 in service, most of which are bivalent (multi-fuel capable, may be fueled by natural gas or gasoline).

Methanol may be obtained from natural gas, crude oil, coal, or biomass, and transported and pumped in the same way as any other liquid. It does not offer any significant environmental advantages. Because methanol is knock resistant, engines may operate with higher compression ratios and therefore lower fuel consumption. However, methanol is poisonous, and contact (e.g., during refueling) may prove hazardous.

Biofuels may be obtained from agricultural products but at the expense of considerable land area. As far as carbon dioxide is concerned, they are at least theoretically neutral. However, fertilizer required to produce biomass crops emits nitrous oxides, with an adverse effect on global air quality ("greenhouse effect").

In 1975, the Brazilian government and its domestic auto industry initiated a national "Proalcool" program, less from environmental than from economic and political considerations. Instead of selling its surplus sugar on the world market at ridiculously low prices, the intention was to produce ethanol from sugar cane and mix it with gasoline, which, due to the 1973–1974 energy crisis, had also become an expensive commodity in Brazil.

After gasoline engines were adapted to burn the alternative fuel in 1979, the portion of alcohol-fueled cars skyrocketed. By 1985, these accounted for 96 percent of all cars registered in Brazil. Beneficial side effects included the environmental friendliness of the fuel, but its enormous appetite for agricultural land was alarming. As the price of sugar on the world market rose again, exports increased. Alcohol fuels became scarce and soon were priced almost on par with unblended gasoline. From 1988, drivers and the auto industry again switched to pure gasoline operation. The only large-scale conversion to alternative fuels to date may therefore be regarded as a failure.

The editor-in-chief of *La France Automobile*, Louis Baudry de Saunier, in his book *L'automobile* (Paris, 1899), wrote "The gasoline-powered vehicle heads the list of all types of motorcars in service today, and it would appear that it will not be easily displaced from its dominant position."

This is equally true today, more than a century later. Was the century wasted? The truth is that even 25 years after the first energy crisis, traffic on land, on water, and in the air is almost 100 percent dependent on crude oil—a disturbing situation, to say the least.

Remarkably, it seems that politicians, the economy, and the public pay scant attention to machinery, transportation systems, and vehicles that do not happen to use roadways. While road vehicles must meet tough emissions, consumption, and noise standards, and emissions limits are also applied to industry and homes, the so-called off-road sector is free to produce pollutants and noise more or less without constraints. The off-road sector includes lawnmowers, chain saws, rototillers, air compressors, mower-threshers, construction and earth-moving machinery, locomotives, ships, and aircraft. The power sources in question range from two- and four-stroke gasoline engines to a huge variety of diesels to aircraft powerplants. They make a significant contribution to mobile source emissions, including the automobile. Depending on the pollutant, they may account for up to 50 percent of emissions (*Automobile Revue*, Bern, June 26, 1997).

The Automobile Industry and Automotive Technology, 1980 to 2000

Driven by pressure from energy crises and emissions, fuel economy, and safety standards, the last quarter of the previous century shows the beginnings of motor vehicle technology more in harmony with its environment. The reciprocating piston engine, whose basic principles were defined by Maybach around 1885, underwent remarkable development. Given that various petroleum products are able to meet the demands posed by advanced engine technology, piston engines continue to offer optimization possibilities and inherent advantages. Today, fuels and lubricants have taken on the aspects of design components.

The introduction of catalytic converters (the European Union made catalytic converters obligatory for all models, as of January 1, 1993) added to design and manufacturing complexity, weight, consumption, and cost, even as engine output dropped. To restore the status quo in terms of engine efficiency, as well as working toward meeting future regulations and the demands of market competition, auto makers undertook an entire series of design measures.

In 1975, Honda introduced a stratified charge engine that met the U.S. emissions standards of that time without post-treatment of the exhaust (i.e., without resorting to a thermal reactor or catalytic converter). The CVCC engine (Figure 207) had two intake valves next to its single exhaust valve per cylinder: a normally sized intake valve to supply a lean mixture in the combustion chamber, and a smaller valve with a prechamber for a rich mixture near the spark plug. Honda was able to meet standards for carbon monoxide (CO) and oxides of nitrogen (NO_x), but not for hydrocarbon emissions (HC), and eventually had to resort to a catalytic converter.

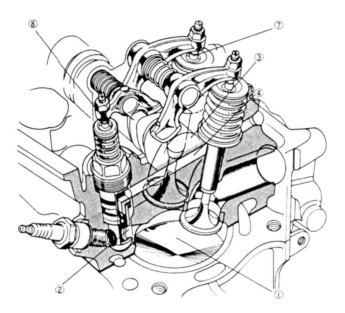

Figure 207. Honda CVCC stratified charge engine, 1975. Stratified charge was addressed as early as Otto's patent, number 532 of 1877. Charge stratification is achieved with the help of a prechamber, in Honda's case fitted with an intake valve. A rich mixture is formed near the spark plug, while the mixture is progressively leaner in the direction of the cylinder wall or piston. CVCC is an acronym for Compound Vortex Controlled Combustion. 1. Combustion chamber; 2. Prechamber; 3. Intake port to main combustion chamber; 4. Intake passage for prechamber; 7. Main intake valve; and 8. Prechamber intake valve.

247

Unleaded fuel is a prerequisite for catalytic converters; lead additives would quickly destroy catalysts. In the carburetor era, unleaded fuel was not available. Oil firms added tetraethyl lead to gasoline to raise fuel octane number (giving it less tendency to knock). Governments around the world responded to concerns regarding environmental hazards. In 1970, U.S. President Richard M. Nixon addressed the topic in his "Reorganization Plan Number 3," submitted to Congress on July 9 of that year, which resulted in the formation of the Environmental Protection Agency in December 1970. As of January 1, 1972, the West German government limited the lead content of motor fuel to 0.40 grams/liter, to 0.15 grams/liter from January 1, 1976, and completely free of lead (regular grade gasoline) as of January 1, 1988. The intervening years saw considerable debate between the government, on one hand, and the auto and oil industries on the other.

Twenty and thirty years ago, lead, benzol, bromide, and chloride fuel additives stood in the way of clean combustion. As we enter the next millennium, the culprit is the sulfur content of gasoline and diesel fuels. Because of catalytic converters downstream of the engine, gasoline powerplants depend on a rich air/fuel mixture. If sulfur were removed from fuel, these engines could operate with a leaner mixture. This will, however, necessitate NO_x adsorbers, de-NO_x, or similar catalytic converter designs (see Appendix), which convert oxides of nitrogen into harmless substances. The combination of lean mixtures and direct gasoline injection, already practiced on the Japanese market by Mitsubishi and Toyota, promises fuel economy increases of approximately 15 percent. For diesel engines, lower sulfur content will reduce particulate emissions.

Further improvements in the field of engine mechanicals include supercharging, hydraulic valve lifters, multi-valve technology, variable valve timing, alloy engines, engines with three or fewer cylinders, Miller cycle, balance shafts, reshaped combustion chambers, variable intake manifolds, cylinder shutoff, and other measures. These improvements are described in detail in the following paragraphs.

By precompressing air (or the air/fuel mixture) before it enters the cylinders (i.e., supercharging), the power output of any given size engine can be increased. To date, all three supercharging methods have been applied to automotive engine design: mechanical supercharging (pistons, screw compressors, vaned rotors), exhaust-driven supercharging (combined exhaust turbine and air compressor, see Appendix), and pressure-wave supercharging (energy transfer between exhaust gas and intake air). The Roots blower, invented in 1848, saw its first large-scale (automotive) applications in the 1920s, as it was used to increase the power output of sports cars. Today, mechanical, turbo, and pressure wave supercharging are employed on diesel and gasoline engines to increase torque at low engine speeds, often in combination with charge-air intercoolers.

Hydraulic valve lifters were first used to reduce engine noise (Yellow Coach, 1932), and then to reduce maintenance (Buick, 1949). The first German car to employ hydraulic lifters was the Opel Kapitän/Admiral of 1964. Hydraulic lifters ensure lash-free operation and compensate for dimensional changes in the valve train. Hydraulic lifters are virtually mandatory on modern multi-valve engines because they eliminate the need to check and adjust valve clearances.

Poppet valves, used almost universally in modern four-valve engines, date back to James Watt's experimental steam engine of 1780. For approximately 100 years, production cars used one intake and one exhaust valve per cylinder. From the mid-1980s, three or more valves per cylinder began to appear. Beginning in 1906, the classic goals of multi-valve (usually four-valve) technology as applied to racing engine design were increased power through better volumetric efficiency, higher engine speeds, and lower mechanical stresses due to reduced valvetrain masses. Today's production car goals, however, are greater freedom in shaping the combustion chamber, especially with regard to its compactness, heat transfer, and gas flow; a central spark plug location; and intake port shutoff and variable valve timing and variable intake systems to achieve high mean effective pressure even at low engine speeds, which translates into lower fuel consumption and more desirable exhaust gas characteristics.

The shift from performance-oriented to exhaust-emissions-oriented four-valve engines is probably marked by the Jensen-Healey of 1972, powered by a Lotus DOHC four-cylinder engine (Figure 208). Jensen and Lotus modified its compression ratio, camshaft, and carburetor system to meet the exhaust standards of the time, allowing the car to be exported to the United States. In 1982, Honda followed with the ET three-valve engine. The following year, Saab, Mercedes/Cosworth, Jaguar, Toyota, BMW, Volkswagen, Audi, and others joined the club, most with four-valve engines. In 1985, Maserati exhibited a six-valve (per cylinder) engine; Mitsubishi presented the first production five-valve engine in 1989. Today, most major auto makers offer three-, four-, or five-valve engines in their product lines.

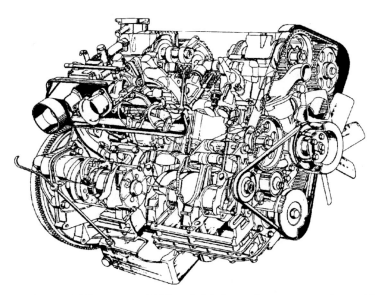

Figure 208. Lotus 907 engine, 1973. A high-performance engine with optimum exhaust emissions for its time: aluminum block, double overhead cams, four valves per cylinder, twin Dell'Orto sidedraft two-barrel carburetors (or Stromberg carburetors for the U.S. market), and timing belt. Intended for the U.S.-market Jensen-Healey. Four-cylinder engine, 1972 cc, 142 hp (105 kW).

For a long time, OHC and multi-valve engines were reserved for racing and sports cars because of the cost of locating the cam above the valves, and driving it by means of a shaft and sets of bevel gears. Since the early 1950s, roller chains (Mercedes and Alfa Romeo) have been employed to drive overhead valves in passenger car applications, joined by rubber cog belts in 1962 (Glas). An ongoing objective is to replace the mechanical camshaft with an "electronic camshaft." Researchers have been working on this concept for at least 25 years. These systems consist of one computer-controlled hydraulic actuator per valve.

With individual valve control, timing and valve lift can be dynamically modified in response to load and rpm conditions. The throttle, bane of the gasoline engine, could be replaced by a switch that meters the charge per cylinder. It also can close the valves earlier or later, changing the effective compression ratio (Miller cycle). Theoretically, it would even be possible to omit reverse gear, because such an engine could rotate in either direction. Another advantage is the ability to run the engine as a compressor under braking, largely taking over the function of the service brakes of the vehicle. The engine could shut down when the vehicle is stationary. For driving from a standstill, a push on the accelerator pedal restarts the engine. Moreover, cylinders can be shut down when only minimal power is needed.

Such an engine promises to deliver high fuel economy, low emissions, and low weight, while providing more power and torque. Although there are currently no production cars with "electronic camshafts," partial solutions have already been achieved—but not without significantly higher design and manufacturing effort.

In 1985, Alfa Romeo introduced an electronically controlled variable valve timing, which introduced a phase shift into the intake camshaft timing. Honda followed in 1988 with variable valve timing and lift for both intake and exhaust valves (VTEC, Figure 209). The Mercedes S-Class of 1998 was equipped with cylinder shutoff as a fuel-saving measure. In operation, the shutdown system deactivates the valves and fuel injection for cylinders 2 and 3 on the right bank, and 5 and 8 on the left cylinder bank, to ensure evenly spaced firing pulses.

After the second energy crisis, Audi and Volkswagen introduced their "Formel E" models in 1980 and 1981, incorporating engine shutdown and other modifications. Volkswagen made a second attempt with its Golf Ecomatic of 1993, employing a freewheeling automatic transmission. In both cases, the engine shut itself off when the car was stationary and restarted with a touch of the accelerator pedal. Also, in both cases, production soon was halted because of lack of consumer interest. In 1995, Mazda introduced the Xedos 9, with its Miller cycle engine (Mazda Millennia S in the United States). Its manufacturer claims higher thermal efficiency and particularly low fuel consumption at part throttle.

The engine of the Volkswagen Lupo 3L TDI of 1999 represents a compendium of modern engine technology, an advancement over the Japanese high-performance microcars in the sub-one-liter class, and quite likely the first fully functional production car offering fuel consumption of approximately 3 liters per 100 km (78 mpg).

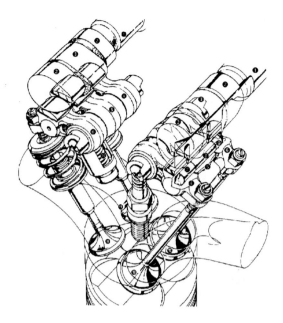

Figure 209. Honda VTEC valve train, 1988. A four-cylinder four-valve engine with double overhead cams, each with three lobes per cylinder; the outer lobes activate the two intake or exhaust valves via short fingers. Small valve overlap is used for low rpm. The center lobes give both valves greater lift and greater overlap for high rpm. The electronic engine management system controls valve lift and timing hydraulically.

Similar to many Japanese microcars of around 1990, the Lupo is powered by a three-cylinder engine—in this case, an aluminum alloy diesel. The three-cylinder layout results in lower friction losses, larger individual combustion chamber volumes, compact exterior dimensions, and an attractive torque curve. However, it requires a counter-rotating balance shaft to compensate for the oscillating and rotating masses of crankshaft and pistons, as well as gas forces.

The overhead camshaft of the Lupo is driven by a timing belt and has three lobes per cylinder. The third lobe actuates a unit injector, which is aimed directly into the cylinder. A variable-geometry turbocharger and intercooling improve volumetric efficiency, while gas flow and fuel injection are electronically controlled. Thanks to additional fine-tuning of the design, the Lupo powerplant already meets Euro 4 emissions standards, which take effect in 2005. Its specific output (61 hp from 1196 cc) is 51 hp/liter—low compared to the 117 hp/liter of the 1989 Mitsubishi Minica Dangan ZZ (64 hp, 548 cc), which was not sold in Europe. The main goal of the Lupo was not power, but rather fuel economy and emissions reduction, supported by the automatic stop/start system and an automatic transmission designed for economy, with the option of manual shifting for sportier driving.

With body and chassis components of aluminum alloy, the Lupo 3L TDI (Figure 210) weighs 860 kg (1896 lb) against 950–1055 kg (2094–2326 lb) of the Lupo's with different engines.

For comparison, the 1953 Volkswagen Beetle tipped the scales at 730 kg (1609 lb), and the 1989 Mitsubishi Dangan at approximately 600 kg (1323 lb). With a drag coefficient (c_d) of 0.29, the Lupo proves that even small cars can have good aerodynamics. Narrow, low-rolling-resistance tires contribute to its low drag.

Figure 210. Volkswagen Lupo 3L TDI, 1999. Composite body construction results in approximately 50 kg (110 lb) of weight savings. Front fenders, doors, and engine lid are made of light alloy, the rear hatch consists of an aluminum outer skin over a magnesium inner panel, seatback frames and firewall are made of aluminum, and load-carrying structures are made of steel sheet metal. Three-cylinder direct-injection diesel, 1196 cc, 61 hp (45 kW).

The tire industry has been working on low rolling resistance for slightly less than a decade. Such tire designs must meet conflicting goals—low rolling resistance on one hand, and long life and good grip under wet conditions on the other. To maintain the expected qualities while achieving low rolling resistance, special rubber compounds with a high percentage of natural rubber, a medium silica content (long molecular chains), a new type of carbon black, and Kevlar carcasses were developed for these tires. They generate approximately 10 percent lower rolling resistance at the tread.

Subaru introduced all-wheel drive for pickups, station wagons, and passenger cars in 1978 (Figure 167), followed by AMC (Eagle, 1979), Audi (Quattro, 1980), and others. Four-wheel drive finds its main application in leisure and off-road vehicles. Increasingly, passenger cars, with their greater need for low weight, are being offered with vehicle dynamic control systems. Although these cannot match the traction of four-wheel drive, they can counteract vehicle spin tendencies. Sensors report to a central control unit, where steering wheel input angles are compared to experimentally determined values. Almost instantly, in the event of imminent loss of control, the system brakes individual wheels as needed, and the electronic engine management system reduces power.

In its function and design, vehicle dynamic control is a further development of anti-lock brake systems (ABS) and anti-slip regulation (ASR). It stabilizes the vehicle, but not only during braking and acceleration. It performs its functions even while the vehicle is rolling freely (i.e., in any driving situation). The first passenger car with vehicle dynamic control was Cadillac, with its integrated chassis control system (ICCS) and the Mercedes S-Class with its electronic stability program (ESP), both in 1995. The concept, which was originally intended to improve safety, is now also used to address problematical chassis designs, such as the Mercedes A-Class (1998), Smart (1999), and Audi TT (1999–2000).

Modern anti-lock brakes trace their ancestry to research by Fritz Ostwald. In 1940, he designed and built an electro-pneumatic brake controller that reduced slip between tire and road, allowing only short intervals of skidding, between which the wheel rolled under braking. The first car with ABS, now regulated electronically, was the Lincoln Mark III of 1968 (rear wheels only) and later the BMW 745i and Mercedes 450 SEL (Bosch, select-low), both in 1978. Beginning in 1982, Audi, Opel, Honda, and others also offered ABS, initially as an added-cost option. Today, there is hardly a European, American, or Japanese passenger car without ABS which, similar to ESP, represents a substantial improvement in passenger car safety.

Four-wheel steering, promoted by Honda, Mazda, and Nissan as of 1987, does not appear to fill a genuine need in the passenger car sector. Except for BMW AG, which offered its 850i with optional rear-wheel steer from 1991, European manufacturers have always preferred "passive" rear-steering suspensions incorporating elastokinematic elements. Large, asymmetrical rubber elements permit slight shifting and angularity of the wheel and axle. The resulting steering effect counters the tendency of the rear end to come around at high cornering speeds, thereby increasing dynamic safety. Costs, design considerations, and weight are far lower than the active-steer systems favored by Japanese makers. The pioneer in passive rear-wheel steering was the Porsche 928 of 1977, followed by Mercedes, Volkswagen, Opel, and others.

Two transverse arms and a coil spring, or, more recently, a single transverse arm, MacPherson strut, and stabilizer bar, with outer pickup points located "inside" the wheel, are state of the art for front suspensions. If transverse arms are each divided into two rods, whose axes intersect outside the wheel, the steering offset will remain constant under all conditions. The steering axis is virtual because it is no longer formed by the outer pickup points. As a result, external forces and drive forces can largely be isolated from the steering system, even under heavy acceleration (front-wheel drive), making the car track more accurately and reducing tire wear. Furthermore, more space inside the wheel accommodates the brake. This five-link suspension, proposed by Fritz Ostwald in 1958, is also suitable for rear axles. It was first implemented on the 1994 Audi A8 (Figure 211).

Despite all its advantages—cost-effective manufacturing, increased rigidity, torsional stiffness, crash safety, and lower weight—unit body construction provided car owners with a new set of problems. With unit bodies, owners would have to combat rust not only on nonstructural

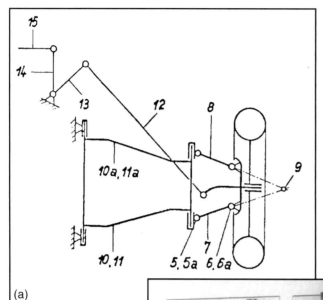

Figure 211. Audi five-link suspension. (a) The transverse arm is divided into two links (7 and 8); their axes meet outside the wheel (at 9), to Ostwald's 1958 German patent number 1,077,058. (b) Audi suspension, here the 1995 Audi A6. Two upper and two lower rods each with their own pickup points and a virtual axes of rotation. The shock strut no longer performs any locating function.

parts such as doors and fenders, but also on structural members. Unit body maintenance and repair costs were higher than for the old body-on-frame method, especially because auto manufacturers held back on anti-corrosion measures until well into the 1970s.

The first manufacturer to employ full-body galvanizing on a unit-body car was Porsche, in 1976, on its limited-volume 911 model; Audi followed in larger numbers in 1985. The entire body structure, including the floorpan, consisted of steel galvanized on both sides, with corresponding positive effects on the cost of ownership, vehicle life, warranty duration, and resale value. Immersion primer, undercoating, and hot wax continued to be applied as in the past. Other manufacturers followed Audi's lead. Corrosion of structural and nonstructural parts—at least among high-end cars—is no longer a problem. Six- to twelve-year warranties against rust-through are customary.

A departure from unit bodywork in the conventional sense is represented by the Audi Space Frame (ASF) concept of 1993–1994, consisting of a load-carrying structure and a partially stressed shell. The aluminum space frame consists of closed extrusions; formed sheet aluminum panels such as trunk walls, floorpan, and roof are welded, bonded, clamped, or clinched to the space frame. The aluminum bodywork is lighter and more corrosion resistant than a comparable steel body.

Passenger cars have employed aluminum fenders, lids, and other body and chassis components since around 1900. Even entire bodies were sheathed in aluminum panels (Figure 212) or formed of aluminum sheet, while conventional materials were retained for frames. Usually, these were individual creations, demonstrators, sports, or luxury cars built in extremely limited numbers. Manufacturing aluminum bodywork on the assembly line, at a reasonable cost, posed an insurmountable problem at the time. A case in point is the SA Panhard & Levassor 1947 Dyna, a small car that ventured to use aluminum alloy for various components including body parts, firewall, and floorpan, but reverted to steel in 1958. With its model year 2000 A2 (Figure 213), Audi is the first car maker to offer a volume-production aluminum automobile.

Apart from materials choices, modern cars offer greater crash safety margins, in offset and frontal crashes, side impact, rear-end collisions, and rollovers. Since 1983, independent crash

Figure 212. Protos 26/50 HP, 1911. Nine-seat tourer built for the Sultan of Langkat (on the island of Sumatra, Indonesia), with aluminum skin over a wooden framework by J.W. Utermöhle, Carosserie-Werke Köln und Berlin. A clever design solution for the side panels, each with three doors and pillars in the seat box walls, lends an attractive and graceful air to this heavy, long-wheelbase car with duplex rear tires. Six-cylinder engine, 6840 cc, 50 hp (37 kW).

Figure 213. Audi A2, 2000. Compact five-seat sedan with second generation ASF aluminum body and space frame. First all-aluminum volume production car. Weight 895 kg (1973 lb) with gasoline engine or 990 kg (2183 lb) with diesel, aerodynamic drag coefficient (c_d) of 0.28. Four-cylinder gasoline engine, 1390 cc, 75 hp (55 kW).

tests performed by auto clubs and magazines have contributed to the cause of safety by making the car-buying public more aware of a topic that was once brushed aside. Today, even compact cars offer low risk of injury, even in 40 mph crashes with 40 percent offset performed according to the Euro-NCAP (New Car Assessment Program). Mutually optimized front chassis structures and compatibility issues between light and heavy cars require further development.

After the unfortunate American ESV intermezzo in the early 1970s, which produced more questions than answers, the industry eventually reached a more reality-based view of the entire safety issue, as Barényi had already proposed in the 1940s. Aside from the car body, with its defined crumple zones and side impact protection, the location of the steering box and fuel tank, design of the steering column, steering wheel, and instrument panel, seat mounting, windshield wipers, and other components and assemblies had to be taken into account. In these areas, considerable progress has been made since the late 1970s, especially in the area of restraint systems. Airbags, safety belts (Figure 214), belt tensioners, and belt force limiters are regarded as one interdependent safety system, optimized to work synergistically and integrated into the vehicle.

Thirty years ago, it was feared that application of a full suite of passive safety measures would make a car unaffordable. This has not happened. According to a study by Deutsche Automobil Treuhand (DAT), a German auto industry and market research organization, the average new

Figure 214. Safety belts, 1903. Gustave-Désiré Leveau's idea is almost 100 years old. The Frenchman sought to protect his beloved family from being thrown headfirst out of their car in the event of an accident. Leveau recognized the basic principles of modern seat belt design: the heaviest sections of the human body, the chest and the hips, are protected by wide, snug-fitting leather belts. These belts are anchored to the tall backrests to minimize loads on the shoulders. French patent 331,926 of 1903.

car buyer spent 23,600 Deutsche Marks in 1986. With an average household net income of 3,655 Deutsche Marks per month, this represented approximately 6.5 months of savings. This figure approached American levels. In 1993, the typical American car buyer paid approximately 6.3 months of income for a new car, compared to 5.6 months in 1983. Despite its increased complexity and higher development costs, the modern automobile has become an affordable object for a majority of the population.

It appeared that safety regulations—more precisely, rollover and roof impact tests—would give the traditional convertible no chance of survival. An alternative was offered by the roll bar, patented in the United States by Barényi in 1956, applied to a Fiat 2300 Special in 1963, and introduced in production vehicles by Porsche in 1967. The Porsche 911 Targa (Figure 215) soon spawned an entire series of imitators, including the Porsche 914 (Volkswagen-Porsche outside North American markets) and Triumph Stag (both 1970), 1971 BMW 2002 Cabrio by Baur, 1973 Fiat X1/9 by Bertone, 1976 Opel Kadett Aero by Baur, and 1979 Volkswagen Golf by Karmann, which to this day carries a roll bar. Even in those days, convertible tops consisted of multiple layers and in some cases included heated rear glass to ensure that convertibles could be driven in winter without sacrificing comfort.

Figure 215. Porsche 911 Targa, 1967. Two years after the 911 entered production in September 1964, Porsche introduced the Targa version of its sports car, a convertible-like variation with a racy roll bar clad in brushed stainless steel. The rigid roof could be removed and stowed for open-air motoring, while the plastic rear window could be unzipped. Later models had a glass backlight. Six-cylinder engine, 1991 cc, 110 hp (81 kW).

In the United States, only Chevrolet (Corvette Stingray coupe, 1967) and Ford (Mustang III show car by Ghia, 1976) paid any attention to the Targa bar. In the United States, air conditioning, perceived safety issues, and, in urban areas, pollutant-laden air caused a drastic drop in convertible sales. Convertibles had reached their high-water mark in 1965, with almost 510,000 units sold. However, nine years later, they managed only 28,000 sales—by American standards, too few to justify production. American Motors had abandoned convertible production in 1968; Chrysler followed in 1971, Ford in 1973, and Cadillac, the last holdout, in 1976.

In Europe as well, convertible sales dropped, but production never entirely ceased. As always, Alfa Romeo, BMW, MG, Mercedes, Peugeot, Porsche, Rolls-Royce, and Volkswagen offered open cars in the 1970s. In the early 1980s, these were joined by Fiat, Ford, and, after a short abstinence, Opel. In 1989, Mercedes introduced a roll bar that deployed automatically in emergency situations; BMW followed with two hoops that rose behind the rear seats, and a roll bar integrated in the windshield. Audi moved the hoops to the seat headrests. Peugeot remembered its 1936 "Coach décapotable" and introduced its 206 Cabrio (Figure 216) in autumn 2000. Similar to the Mercedes SLK (1996), its metal roof lowers into the trunk area electrohydraulically. With all of these features, the convertible has achieved a new standard of quality: it is safer than ever and has recaptured the elegance it enjoyed before the arrival of the Targa concept.

Figure 216. Peugeot 206 Coupé Cabriolet, 2000. While the rear section and trunk of the 1936 Peugeot Coach Eclipse were unusually long to provide stowage for the single-piece roof, its modern successor is served by a short, tall rear section. The rigid roof folds in half and disappears at the touch of a button.

Pickup trucks are a uniquely American phenomenon, built on a truck or delivery van chassis, or even employing passenger car suspension. Over the years, they have evolved their own unique appearance. They are fitted with one or, in the case of extended cabs, two rows of seats, have an open cargo box with stylistically integrated, non-folding sides, and offer ride comfort and equipment comparable to passenger cars. Because pickups can be used as workhorses during the week (Figure 217), shopping carts on Friday, off-road vehicles on Sunday, and, with a camper shell fitted to their cargo beds, as a rolling motel for vacations, production numbers rose, in part also because of the introduction of four-wheel drive in 1973. Meanwhile, pickups have become more popular than sedans. In 1999, the Ford F-Series and Chevrolet Silverado pickups were the best-selling vehicles in the United States, with the Dodge Ram and Ford Ranger in fifth and seventh places, respectively. In other words, pickups accounted for four of the top ten selling vehicles in the U.S. market.

Figure 217. GMC Pickup 250-24, 1954. A pickup built by the General Motors Truck and Coach Division, in the style of the times with a chrome-plated wide-mouth grille, single-piece windshield, and running boards. This one-ton truck, here with an after-market corrugated shell installed in the box, was equipped with a General Motors Hydramatic automatic transmission. Six-cylinder engine, 4072 cc, 125 hp (92 kW).

Another American phenomenon is the popularity of sport utility vehicles (SUVs), a hybrid of off-road vehicle, van, and family sedan. They trace their roots to the 1970 Range Rover, the first off-road vehicle acceptable in polite society, and the Jeep Cherokee/Wagoneer (Figure 218) produced since May 1984. Measuring 4.20 × 1.78 × 1.63 meters (13.8 × 5.84 × 5.35 ft) (length × width × height), it was 53.5 cm (21.1 in.) shorter, 14 cm (5.51 in.) narrower, and 5.5 cm (2.2 in.) lower than previous Cherokee/Wagoneer models, which for their part had grown out of the Jeepster (Figure 138). With these dimensions, the new Jeep model line had essentially achieved passenger car dimensions, except for an extra 30 cm (12 in.) of height, and served as the model for American, Japanese, European, and, most recently, German competitors (Mercedes M-Class, BMW X5, and Audi Allroad).

Simultaneously, the past few years have seen the arrival of the Ford Expedition and Excursion, Lincoln Navigator, and Cadillac Escalade. These vehicles, resembling small buses, are luxuriously equipped and have powerful engines, overall lengths of up to 5.76 meters (18.9 ft.),

Figure 218. Jeep Cherokee 4×4, 1988. Because a more compact vehicle with a lower center of gravity is less likely than a larger vehicle to tip over in off-road operations, even in unskilled hands, Jeep Corporation, then a division of AMC, reduced the exterior dimensions of its multi-purpose 4×4 line. Produced from 1984, the Cherokee/ Wagoneer line marked the starting point of sport utility vehicles (SUVs) of today. Six-cylinder gasoline engine, 3962 cc, 179 hp (132 kW).

widths up to 2.03 meters (6.7 ft), and heights up to 2.02 meters (6.6 ft) tall. Their successors will in all likelihood be offered not only as closed-body SUVs but also fitted with four-seat cabs and enclosed cargo beds, which, with seatbacks folded forward, provide a flat load area approximately 1.83 meters (6 ft) long. From a European and environmental perspective, the question arises whether such luxurious, gargantuan offshoots of the pickup and SUV make any sense.

Similar vehicles have been available since the mid-1980s. Starting with a 1977 Lamborghini off-road truck designed for paramilitary desert operations, the California firm Mobility Technology International (MTI) developed a high-performance all-terrain vehicle, the LM 002, with a luxuriously appointed cabin for four persons and a small cargo bed. Until production ceased in 1991, perhaps 400 were built. Of these, most were sold in the Middle East, and some were exported to the United States.

There, in 1984, the Pentagon had issued a specification for a vehicle to replace the aging Jeep (Figure 137). The future military vehicle was to have a payload of up to 2500 lb (1134 kg). By comparison, the Jeep could carry 800 lb (363 kg). Fuel consumption was to be no more than 9.05 liters/100 km (26 mpg), compared to the Jeep's 12.4 l/100 km (19 mpg). One chassis was to serve several body variations. (The Jeep needed three different chassis.) Of the three candidates, the Jeep's manufacturer, American Motors subsidiary AM General, won the contract to build the High Mobility Multipurpose Wheeled Vehicle—the HMMWV, more readily known as the Humvee and later the Hummer (Figure 219).

Figure 219. Hummer, 1984. Military requirements for rollover resistance, low silhouette, and large ground clearance could be achieved only with a wide track (183 cm [72 in.]), a drivetrain that intrudes into the cabin, and reduction gears at the hubs. This allowed halfshafts and suspension to be mounted higher. More than 100,000 examples of the Hummer have been produced to date. Shown here is a hardtop version with a cargo bed and short cab. V8 diesel engine, 6217 cc, 152 hp (112 kW).

The Hummer, which is suspiciously similar to Lamborghini's 4×4, seats four, but its cabin is executed in a much more spartan style. It, too, has a cargo bed at the rear. Beginning in 1984, tens of thousands were sold to the military without attracting any public interest whatsoever—until news reports from the 1991 Gulf War made it a household name throughout the world. In response to strong public demand, the manufacturer began offering civilian versions in June 1991, a move that had not been planned originally. After all, a vehicle measuring 2.16 meters (7 ft) wide, a high-mounted drivetrain providing cramped space for only four occupants, and an all-up weight of 4600 kg (10,141 lb) is not particularly suited for private use and, in some countries, requires licensing exemptions and a truck driver's license. Meanwhile, AM General has become part of General Motors. Elsewhere, Toyota and URO of Spain were attracted to the Hummer concept and are producing more or less authentic copies.

Monstrous SUVs aside, there was also movement in the other direction. In 1997, Renault introduced a single-box vehicle, the Scénic, based on its compact Mégane, and established an entirely new class of minivans in Europe. The minivan is marked by its single-volume interior, variable seating, and tall body structure (160 cm [63 in.] in the case of the Scénic), on a relatively short chassis (413 cm [162.6 in.]). The idea was by no means new; Mitsubishi (Toppo) and Suzuki (Wagon R) had already offered minis with heights of 177 and 164 cm (69.7 and 64.6 in.), respectively, with an overall length of only 329 cm (129.5 in.), but in both cases with a clearly defined, sloping hoodline. These Japanese models themselves drew their inspiration from a predecessor that never left the starting blocks. In 1965, the autonova design group, with a team consisting of Fritz B. Busch, Piu Manzú, and Michael Conrad, displayed their single-box car, the fam (Figure 220) at the IAA (Frankfurt Auto Show). This preceded the Mitsubishi Super Space Wagon of 1979 (Figure 168), scaled 160 cm (63 in.) tall and 340 cm (134 in.) in length, and had four doors and four flip-down seats mounted on a flat cargo floor.

Figure 220. Autonova fam, 1965. With assistance from NSU, Glas, Recaro, VDO, and Boge, this early minivan featured a variable interior configuration and a horizontally split rear hatch. Unusual features included an electronically controlled manual trans-mission, load-leveling suspension struts, and progressive steering, which reduced the steering wheel travel to 280 degrees lock-to-lock. Instead of a steering wheel, it used a yoke with integral switches. It still employed rear-wheel drive. Bodywork by Sibona & Basano of Turin. Four-cylinder engine by Glas, 1281 cc, 60 hp (44 kW).

Today, to operate successfully in the market, it is no longer sufficient to offer only one body style in only one vehicle class, or even to offer only a conventional notchback sedan, as was the case recently. Remaining with Renault as an example, the Mégane is offered as a fast-back and notchback sedan, as a station wagon, coupe, convertible, and minivan. Below and above, other model lines, each with its own variations, extend the selection, making the customer's model choice even more difficult. Other manufacturers employ similar strategies.

Shifting model policies are also an expression of the current economic and political situation, resulting from the collapse of the Soviet Bloc in 1989–1990 and subsequent globalization. Before this watershed event, the European auto industry found itself in a decidedly inferior position relative to its Japanese competitors, specifically in terms of financial power, productivity, and flexibility (i.e., agility in model policies). The industry realized it was incapable of tailoring individual models to customers' wishes and unable to capture narrow market niches while operating profitably. In 1988–1989, European firms needed an average of 35.9 hours to produce a single car; Japanese plants in Japan needed 19.1 hours, and Japanese transplants in the United States needed 19.6 hours. In the 1980s, Toyota and Nissan had cash flow averaging 1.7 times capital investment; Volkswagen, Fiat, and Peugeot cash flow just managed to match investment.

The fall of the Berlin Wall in 1989 and the dissolution of the previous Soviet Bloc trading alliance in 1991 (the Council for Mutual Economic Assistance [COMECON]) was followed by a wave of West German corporate investment in eastern Europe—and beyond, amplified by structural weakness in West Germany's own economy. At home, West German firms found themselves surrounded by obstacles—excessive taxes and fees, high wages, short work weeks, underutilization of machinery, excessive government investment in industry, bureaucratic harassment, and subsidies to maintain old industries (e.g., coal and shipyards). Innovation stagnated, as evidenced by the decreasing number of patent applications, particularly in future growth sectors. More serious, however, is the lack of readiness on the part of industry and the economy to capitalize on existing inventions and patents.

To at least reinforce the rather shaky concept of "made in Germany," at least in the auto sector, the automotive industry subjected itself to a radical procedure. Modeled in part on the Japanese "lean production" concept, the industry reduced its manufacturing depth by outsourcing or purchasing, forced design and production to greater cooperation, reduced the number of parts required (modular assembly concepts), adjusted working hours to demand, improved quality during the manufacturing process, introduced just-in-time delivery, relocated supplier industries directly at the manufacturing plant (Mercedes' Smart car), and cut staff in production, administration, and management.

These measures began to take effect. While German auto plants (with the exception of BMW and the American subsidiaries Ford and Opel) continued to operate "in the red" in 1993, they returned to profitability beginning in 1994–1995. One contributory cause was a model offensive that introduced—among others—vans, roadsters, and off-road vehicles into market niches once held by the Japanese. Moreover, with lightweight alloys, direct-injection diesels, active vehicle dynamic control, and other improvements, they add credibility to German industry's claim to technological leadership.

Even so, at this time German manufacturers had a strategic disadvantage compared to their American and Japanese competitors because they had no manufacturing presence in those markets and instead operated only as importers. As a result, they were vulnerable to market swings, exchange rate fluctuations, and import restrictions.

To reduce exposure to these risks, at least in North and South America, the eastward expansion of the German auto industry (e.g., new Opel plant at Gleiwitz, Poland; new Audi plant at Györ, Hungary) was soon followed by a similar move westward. BMW established its

Spartanburg, South Carolina, plant in 1994; Mercedes its Tuscaloosa, Alabama, factory in 1997; and Mercedes and Audi opened new manufacturing facilities in Brazil. Volkswagen has been producing cars in Brazil since 1953 and in Mexico since 1967. A Volkswagen plant in Westmoreland County, Pennsylvania lasted only from 1978 to 1988 (i.e., precisely until the time the Japanese were erecting transplants of their own).

Only after the once politically isolated markets of eastern Europe, Russia, mainland China, and even Vietnam are opened, and the Internet is made available for private commerce, can we speak of a truly global marketplace. Any firm wishing to survive in the global market must have plants and distribution networks in the most important regions, and a complete model lineup ranging from small cars to the luxury class and off-road vehicles. It also must be large enough to command low prices from its suppliers. Companies that do not expand to a viable size soon will become victims of the consolidation process. Not surprisingly, we have recently witnessed a spate of takeovers, mergers, and joint ventures on the largest possible scale. A few of the most important ones are as follows.

Volkswagen took over Seat in 1986, Skoda in 1991, and Bentley in 1998, and secured the rights to the Bugatti name. Also in 1998, Volkswagen subsidiary Audi acquired Italian sports car maker Lamborghini and British engine maker Cosworth. In 1996, Bosch bought the brake system division of American automotive equipment supplier AlliedSignal. Continental acquired the ITT brake and chassis operations in 1998, bringing Teves Frankfurt back under German control. In 1994, BMW acquired Rover, with the rights to the brands MG, Mini, and other marques of erstwhile conglomerate BL Ltd., and, as of 2003, will take over from Volkswagen the manufacturing and distribution of Rolls-Royce. Due to losses, BMW divested itself of Rover and Land Rover in 2000, retaining only Mini. In 1996, the last independent German tractor manufacturer, Fendt, became part of American manufacturer Massey-Ferguson. During the 1990s, Ford (USA) took control of Jaguar, Aston Martin, Volvo, and Land Rover (2000). General Motors did likewise with Saab and British sports car maker Lotus. Finally, Fiat absorbed Alfa Romeo, Lancia, Ferrari, and Maserati. In 2000, General Motors took a 20 percent stake in the sole remaining automobile manufacturing concern in Italy (see Table 7 in the preceding chapter).

However, the most spectacular union of two automotive concerns in the waning days of the twentieth century was the fusion of Mercedes-Benz AG and Chrysler Corporation, creating DaimlerChrysler AG in 1998. The largest German industrial concern had joined with the third-largest American auto manufacturer and, with subsequent mergers of other German firms with foreign entities in other economic sectors, transformed the nation's economy. It no longer appears possible to pursue a course of national industrial policy. We are all playing by a new set of rules, those of a global economic order. In the world of the automobile at least, German firms play an active role. With their newfound strength, they are no longer victims but rather are the movers and shakers of a global race for survival. Looking back to the early 1990s and the erstwhile dominance of the Japanese industry, this is a situation that no one would have dreamed possible.

These unions, especially that of Mercedes and Chrysler, put special pressure on those firms that have heretofore operated on only one or two continents. In the future, it will no longer be sufficient to build a single type of vehicle. Even if these companies concentrate on only their core business, as is the current fashion, they will need to carry not only passenger cars and commercial vehicles in their product lines but also motorcycles and bicycles. These, too, are means of personal transportation and mobility.

The Automobile Industry at the Turn of the Century, II

In the time span from 1900 to 2000, the role of the automobile in the Western world transformed itself from a sporting implement and toy of the aristocracy to a consumer good and indispensable tool of the ordinary citizen. It not only imparts mobility to individuals, but also provides the economic well-being of millions. In the industrialized nations, the automobile has become a decisive economic factor. Quantifiable damage, such as acreage consumed by industry, or accidental traffic deaths (1998: 408,000 worldwide, 101,940 in Europe, 45,030 in North America, and 7,780 in Germany) as well as hidden negative impact on the quality of life, air pollution, and rapacious consumption of resources, are accepted by society.

It would be almost impossible to find a family that does not have a car, regardless of whether they are top earners, middle class, or unemployed. Some families own several vehicles, including motorcycles or vintage cars. Today, even students drive cars of their own, and not necessarily the smallest models. And it is no wonder. During the course of decades, the relationship between new car prices and monthly income has steadily improved, as has been pointed out here several times. In 1991, the average trained worker, earning 4,446 Deutsche Marks per month (married, two children, employed by the city of Munich) needed only 4.5 months' income to buy a new Volkswagen Golf 1.3, at 19,975 Deutsche Marks.

The automobile has become the dominant means of transportation on this planet. The Ifo Institute in Germany estimates that for 1996, passenger cars performed 82 percent of all transportation duties related to moving people, while 66 percent of all freight duties were performed by trucks. Public transportation, both metropolitan and commuter (e.g., trains, buses, subways, and streetcars) accounted for only 14.8 percent of traffic; railways handled only 15.4 percent of freight traffic. By comparison, only a few years earlier in 1982, public railway and bus service had accounted for 19.5 percent of passenger traffic, and the railroads handled 25.7 percent of freight service.

What this means is that public transportation capacity is under-utilized and, furthermore, that despite new construction, the road surface area available to the individual vehicle is shrinking steadily, particularly in central Europe and Japan. To better utilize roads and avoid collapse of the transportation system, traffic systems (see Appendix, Traffic management systems) have been developed.

These may be able to make road traffic more efficient, more environmentally friendly, and safer; however, they cannot reduce traffic volume. With main arterial roads, secondary roads also will be filled to capacity, with the familiar negative consequences.

Since the energy crises, much has been written and said on the topic of "traffic elimination" but with few concrete results. Except for temporary political and economic disturbances, traffic in general and in all its forms has increased from decade to decade. Also, road traffic has "profited" from this general increase by a disproportionate amount, and essentially all modes of transportation are powered by internal combustion engines that rely on fossil fuels and emit noise and poisonous exhaust emissions. Viewed from this perspective, even if partial improvements have been made, the past 115 years have seen no progress in our means of transportation.

115 Years of Motorized Traffic and Transportation

The Road Behind, The Road Ahead

Similar to the steam engine and electricity before it, the Otto-cycle engine and its application to road vehicles have transformed human existence. In the car-making nations, the motor vehicle has grown to become a decisive economic factor. Even seen entirely on its own merits, the auto industry has created considerable value quite independently of its effects on the economy as a whole. Added to this are the supplier industry and the industries that process raw materials. Further economic influences include roadbuilding and the repair trade. In the United States, that portion of net social product traceable to the automobile and its ancillary industries is estimated at approximately 20 percent. In the Federal Republic of Germany, every seventh job is directly or indirectly dependent on the automobile.

Motorization has had special significance in agriculture. Without lightweight combustion engines in agricultural vehicles and implements, it is likely that the world's ever-increasing population could not be fed.

Granted, these societal benefits are accompanied by undesirable side effects. Noise, the Number One environmental problem in major cities, increases the risk of heart attacks. Mechanical shock and acid rain literally undermine the foundations of our cities. Exhaust emissions negatively impact public health and alter climate on a global scale. Combustion of hydrocarbon fuels in liquid or gaseous form, as in reciprocating or rotary piston engines and gas turbines, will produce harmful emissions in accordance with the laws of physics, ultimately making these engines unsuitable for future modes of transportation. Automobiles and commercial vehicles are indispensable elements of civilization. Uncoupled from the engines of today and fitted with environmentally neutral propulsion systems, it may be assumed that they still face a bright future.

While commercial vehicles transport goods from point A to point B and (with the exception of race trucks) strictly represent a means of transportation, the passenger car also provides transfiguration and promise, mobility and security. It is an attribute of modern civilization, enabling the expression of primal urges and the conquest of primal fears.

269

Within the flow of traffic, the automobile instills a feeling of belonging, and, on lonely roads, a sense of adventure, with a guaranteed outcome. The automobile is a castle. Keys or chips replace the drawbridges of old. The growl of the engine takes the place of thundering artillery, to strike fear in the hearts of surrounding traffic—the enemy. From his battlements, the lord of the castle looks down on those who toil around him. The automobile is a cave that no one else can enter and that is defended by all possible means. The automobile is the womb; inside, one is warm, comfortable, and protected. The automobile is at once static—one sits inside. However, and this is unprecedented, it is also dynamic because one moves with the car. It was invented to satisfy mobility, humanity's other primal urge. The automobile allows us to choose between the eternally conflicting forces that drive us: within or without, stability or a nomadic existence, domestication or rebellion.

Even at a standstill, the automobile evokes pleasant emotions—status, power, and individuality. In motion, it conjures up the mastery of time, space, and speed—without requiring us to leave the protection of the capsule. It amplifies the limited abilities of the human body with regard to distance, speed, carrying ability, and capacity. No other human achievement—not houses, ships, coaches, railways, aircraft, or motorcycles—can compete with the automobile in this regard. A government that withholds mobility from its subjects, today synonymous with the automobile, is voted out of office or overthrown. Remember East Germany and the People's Republic of China, the latter of which has learned from the former's mistakes. Alarmists who, at least in cities and urban areas, seek to replace the conventional automobile with a more "responsible" and appropriate automobile or other means of transport, are forgotten as soon as they have committed their thoughts to paper.

With one exception. Swiss business consultant Nikolaus Hayek sought to transfer the success of the Swatch watch to an automobile. However, it was intended as "not simply a new small car, but rather an integrated automobile concept as a contribution to solving our planet's growing traffic problems" (advertising for MC Micro Compact Car AG in the *Süddeutsche Zeitung*, June 23, 1995).

It appears that this advertisement refers to "networking" of all available transportation systems (see Vester, Frederic, *Ausfahrt Zukunft*, "Next Exit, the Future," Munich, 1990), in this case primarily a cooperative effort between road and rail. The auto industry would produce a compact, tall, multi-mode short-range vehicle powered by a quiet and environmentally friendly powerplant. This would be transported from town to town on double-decker railway cars. The compact cars would enter and leave the railway car laterally and park transversely on the lower level. Seating and accommodations for drivers and passengers would be on the upper level. Other aspects of the concept include reduced parking fees for short-term public parking (e.g., at railway stations and airports), as well as reasonably priced rental vehicles for transporting large items.

This "short car" entered the market in 1998 as the Smart (Figure 221), but networking the Smart and other transportation modes is nowhere in sight and indeed cannot be realized in the form outlined here. It is cheaper to rent a car or hire a taxi at the destination railway station. Moreover, a single automobile design will hardly make a significant contribution to solving our planet's traffic problems, even if it is part of a larger system. Amsterdam's Witkar failed, Tulip (Figure 202) has faded from the scene, and the outcome of Toyota's ongoing Crayon project in Nagoya may be regarded with skepticism. All three of these projects encompassed electric vehicles.

As long as oil prices remain low, thereby blocking development and deployment of alternative drive systems; as long as railway management thinks in bureaucratic rather than economic terms; as long as the will to embrace unconventional solutions (Transrapid, people movers) is lacking; and as long as various modes of transportation are subjected to differing fiscal burdens, there can be no hope of achieving rational transportation policy. An all-encompassing policy must be applied across all of Europe, if not worldwide. This presents, above all, a political challenge.

Figure 221. Smart Coupé, 1998. A microcar developed as part of a misguided attempt at networking two different transportation modes. Without its other mode (commuter rail transportation of small cars), the Smart is as isolated as the PSA Tulip (Figure 202). The Smart exhibits problematic handling because of its rear engine and short wheelbase. With only two seats and limited luggage space, it also lacks utility. Under Mercedes leadership, new manufacturing methods were implemented by outside module suppliers located at the manufacturing plant (Hambach, Lorraine, France). Three-cylinder gasoline engine, 599 cc, 45 hp (33 kW) or 55 hp (40 kW).

Appendix

Glossary

Ackermann steering. To negotiate turns, the front wheels pivot around kingpins or suspension pickup points. To avoid scrubbing of the tires, both imaginary extensions of the axles of both front wheels must intersect the extension of the rear axle or rear wheels at a single point. Today, used almost universally on all motorized road vehicles except motorcycles. Invented by Georg Lankensperger of Munich in 1816, patented in Great Britain by Rudolph Ackermann (patent number 4212) in 1818.

Afterburning. Engine external post-treatment to reduce harmful exhaust pollutants. Catalytic reactors: *See* Catalytic converter. Thermal reactors: Air is injected into a reactor (shaped like an exhaust muffler but fitted with thermal insulation). The added air reacts with noxious compounds produced by the combustion process in the engine. Carbon monoxide is converted to carbon dioxide, and for hydrocarbons two hydrogen atoms combine with one oxygen atom to produce water while the carbon is oxidized to form carbon dioxide.

Air engine (Stirling engine). Heat engine with external combustion. The Stirling engine is driven by expansion of heated air or other gases. Two pistons operating in the same cylinder (displacer and working pistons) move in the same direction or in opposite directions at various times in the cycle, alternately forcing air (or gas) into a cooler, where it is compressed, or a heater, where it expands, thereby producing work.

Battery ignition. *See* Ignition systems.

Buggy. Light, single-horse-drawn, usually open vehicle with two (England) or four (United States) tall wheels. In the United States, this included motorized buggies until around 1912, usually with underfloor engines. Today, the term is applied to a type of open recreational vehicle with limited off-road capability (usually based on the Volkswagen Beetle chassis and engine).

Capacitor discharge ignition system. *See* Ignition systems.

Carburetor. *See* Float carburetor; Spray carburetor.

273

Carnot cycle. In combustion engines, an air/fuel mixture, or air alone, is subjected to a thermodynamic cycle (i.e., its temperature, pressure, and volume are changed). In the process, the object is to convert the greatest possible portion of the supplied fuel's heat content into work, with the lowest possible heat losses (e.g., cooling, radiation, and exhaust). The efficiency of the cycle is the ratio of work obtained to the heat supplied. The theoretical cycle described by Sadi Carnot (1796–1832) would have the greatest possible efficiency if gas or air were subjected to two adiabatic processes (without heat transfer) and two isothermal processes (at constant temperature) during the cycle. (Figure 222)

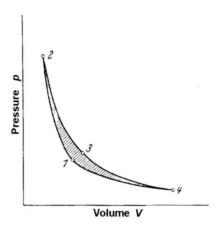

Figure 222. Pressure/Volume (P/V) diagram of the Carnot cycle. The working piston, moving from right to left, compresses air along the isothermal path 4–1 (constant temperature). There is no heat transfer along the adiabatic path 1–2; temperature rises. The highest compression temperature and pressure is achieved at 2, at which point the piston reverses its motion. Fuel supplied along the isothermal path 2–3 ignites in the hot air; temperature does not rise. Thermal energy is converted to work along the adiabatic path 3–4 (cooling through expansion).

Catalytic converter. Catalytic converters, or catalysts, consist of a housing containing carrier material (a substrate) coated with an active substance. Carrier materials may be granulated, sintered monoliths, or metallic. The active coatings, which are sensitive to lead or sulfur, consist of noble metals. Selective (three-way) catalytic converters are used with stoichiometric air/fuel mixtures (approximately 14 kg [31 lb] of air per kg of fuel); NO_x converters are used with lean mixtures (excess air). Catalytic converters convert the noxious exhaust components carbon monoxide (CO), hydrocarbons (HC), and oxides of nitrogen (NO_x) into the harmless substances water (H_2O), carbon dioxide (CO_2), and molecular nitrogen (N_2) (q.v., "Afterburning").

Crank drive. Found on almost all modern piston machines, crank drive has been used to convert reciprocating motion to rotary motion since at least the Middle Ages (e.g., the spinning wheel and a treadle-operated lathe). However, James Watt, who had used cranks on a 1779 model of a two-cylinder steam engine, felt these were not strong enough to transmit significant force. When crank drive in conjunction with steam engines was patented (J. Pickard, patent duration 1780–1795), Watt was forced to find a different solution. Nicolas Cugnot had employed a ratchet-and-pawl mechanism in 1769/1771.

Diesel fuel injection. (Figure 223) Modern diesel fuel injection (also called solid injection) replaced air injection (which employed a compressor and pressure tank) around 1913 for stationary engines, and led to introduction of the diesel engine for vehicle applications in 1922. Several systems came into use for vehicles: prechamber, swirl chamber, and air chamber diesels, each with chambers or reservoirs separate from the combustion chamber and into which fuel was injected or in which stored air performed a "bellows" function. Injection pressures are approximately 80–150 bar (1200–2250 psi). Today, passenger cars and trucks almost universally employ direct injection, in which fuel is introduced into a single-volume (undivided) combustion chamber. Four systems: conventional pumps in inline or (radial) distributor configuration; common rail systems; unit injectors (pump-nozzle units for individual cylinders); and unit pump systems. Currently, injection pressures are as high as 1350 bar (common rail systems) and 2000 bar (unit injectors). Electrical solenoids, currently used in common rail injection systems, will soon be replaced by piezo injectors that will (initially) permit up to six injection pulses per cycle (benefits to emissions, noise, and catalytic converters).

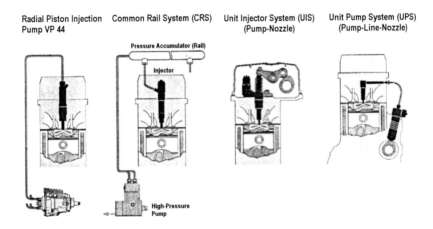

Figure 223. Current diesel direct-injection technology: distributor pump with injector; common rail system with injector; unit injector; and unit pump system.

Direct gear. In a three-speed transmission, third gear is direct, that is, the mainshaft (input shaft) turns at the same speed as the output shaft. The layshaft idles at the same speed.

Distributorless electronic ignition. *See* Ignition systems.

Dry sump lubrication. With dry sump lubrication, the oil supply of an engine is contained not in the crankcase (oil sump) but rather in a separate tank, from which oil is pumped to the lubrication points. The oil flows from these points back to the sump, from where it is scavenged and returned to the tank. Typically used for underfloor and racing engines.

Electric drive. A road vehicle, independent of tracks or electrical power lines, usually fitted with a DC series-wound motor, which supplies its maximum torque the moment it is activated (eliminates the need for change-gear transmission). High acceleration from standstill. Battery-powered electric vehicles exhibit limited range and limited maximum speed.

Electronic ignition. *See* Ignition systems.

Exhaust-driven turbocharger. Exhaust gases drive a turbine wheel, which is mounted coaxially with a compressor wheel. The compressor supplies intake air to the engine cylinders at a slight overpressure (Swiss patent by Alfred J. Büchi, 1905). Originally developed to improve power and efficiency, today turbochargers are more often used to reduce specific fuel consumption. Used on both diesel and gasoline engines.

Fifth-wheel steering. Similar to kingbolt or turntable steering, this is a member of the pivoting-axle family of steering systems (i.e., in negotiating a turn, the entire front axle pivots around a vertical pin or turns on a ring or turntable). The (front) wheels remain parallel to each other. Except for horse-drawn vehicles, these no longer have any application to modern road vehicles for several reasons: with increased steering lock, there is danger of tipping; heavy components result in high frictional losses; the rotation of the entire axle requires a great deal of space; and outside forces acting on the wheels compromise safety and durability of the components. For these reasons, virtually all applications use Ackermann steering (q.v., Ackermann steering).

Flame ignition. *See* Ignition systems.

Float carburetor. (Maybach, 1885; Figure 224) A surface carburetor fitted with a float, c, which moves freely along the feed tube b. The air/fuel mixture is drawn out through the air supply tube e and the funnel-shaped extension of the float, whose constant fuel level ensures a stable air/fuel mixture. (From Sass, *Geschichte des deutschen Verbrennungsmotorenbaues*, Berlin, 1962)

Float carburetor. (Benz, 1886; Figure 225) A surface carburetor with a float, g, which controls fuel supply by means of a needle valve and seat in tube h, and thereby keeps the fuel level constant. Air enters via a trumpet-shaped tube c, mixture is drawn off through tube a. Preheating is accomplished via exhaust gases passing through a lower chamber, at k. (From Sass, *Geschichte des deutschen Verbrennungsmotorenbaues*, Berlin, 1962)

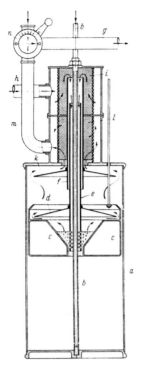

Figure 224. Maybach's float
carburetor, 1885.

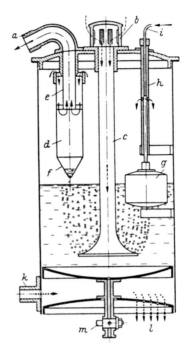

Figure 225. Benz' float
carburetor, 1886.

Flywheel magneto ignition. *See* Ignition systems.

Four-stroke cycle. In the four-stroke engine, the four processes that make up the thermodynamic cycle take place above the piston, in the working cylinder.

First stroke: Intake. As the piston moves downward from top dead center (TDC), a partial
vacuum is created; air streams into the cylinder through the opened intake valve(s). (In the
case of carbureted and indirect-injected engines, air/fuel mixture enters the cylinder.)

Second stroke: Compression. With all valves closed, the air (or air/fuel mixture) in the
cylinder is compressed as the piston moves upward. Direct-injection gasoline engines inject
fuel during this stroke. Pressure and temperature, and therefore the ability of the mixture to do
work, increase.

Third stroke: Power. Just before the piston returns to top dead center (TDC), the mixture of
gasoline engines is ignited by a spark. In the case of diesel engines, fuel injected into the hot,
compressed air ignites. Temperature and pressure rise as a result of combustion, and the
gases in the cylinder expand and drive the piston downward, producing work.

277

Fourth stroke: Exhaust. The exhaust valve(s) opens just before bottom dead center (BDC), allowing burnt gases to expand completely. While returning to top dead center (TDC), the piston forces out the remaining exhaust gases. This is followed by the first stroke as the cycle repeats.

Friction drive. A friction wheel, which can be moved along a line at a right angle to the axis of a flywheel attached to a crankshaft, is capable of transferring torque with a continuously variable ratio from creeping speeds to the top speed of the vehicle. Suitable only for low-powered vehicles (because of slip and wear of the friction surfaces).

Fuel cell. In principle, fuel cells consist of plates separated by membranes. Air, containing oxygen, flows on one side, while hydrogen, ionized into protons and electrons, streams past the other side of the membrane. Protons diffuse through the membrane, producing a negative charge on the hydrogen side and a positive charge on the oxygen side of the membrane. The potential difference (the resulting voltage) is tapped to drive an electric motor. In the process, water is produced and escapes as steam.

Gas turbine. Atmospheric air, after passing through a compressor, enters a combustor. Fuel is sprayed into the combustor and ignited. The gases expand through a turbine, whose output is used (in part) to drive the compressor. The difference between turbine output and power needed to drive the compressor is the useful output of the gas turbine. When installed in vehicles, turbines require auxiliary devices (e.g., heat exchanger or intercooler).

Gasoline fuel injection. Mixture formation by means of a fuel injection pump and injection nozzles, instead of a carburetor. Injection either directly into the combustion chamber or indirectly into the intake manifold upstream of the intake valve(s). Initially mechanically, later electronically controlled. Originally used to increase power and torque; today employed to reduce fuel consumption and emissions.

High-tension magneto ignition. *See* Ignition systems.

Honeycomb radiator. (Maybach, 1900) Further development of tube-core radiator; square tubing (instead of round tubes) with large air flow cross sections and smaller cooling water passages, resulting in higher water flow velocity, reduced weight, and less coolant carried on board.

Hot-tube ignition. *See* Ignition systems.

Hydropneumatic air suspension. Wheel movement is transmitted by means of levers to a piston inside a hollow cylinder, which is partially filled with fluid (oil), partially with gas (usually the oil and gas are separated by a membrane). During load changes, the piston acts on the compressible gas through the incompressible oil. By pumping additional oil into the cylinder, the piston is maintained at its original position. Load leveling is accomplished by means of a ride height controller, valves, oil pump, and oil reservoir.

Ignition systems. *(In chronological order)* Ignition systems ignite the gas or air/fuel mixture inside a combustion engine. Historically, this has been accomplished by various arrangements (see the following). Today, ignition is achieved by means of an electric spark (i.e., a short-duration arc discharge between the electrode and ground of a spark plug).

Trembler coil ignition. (Lenoir, 1860; Figure 226) With the primary circuit closed, a Ruhmkorff induction coil, fed by a galvanic element (voltaic pile) or other source of electricity builds up a magnetic field. When the field reaches a certain strength, it pulls a solenoid which breaks the primary circuit, collapsing the magnetic field. This induces high voltage in the secondary circuit, which is discharged in the form of an electric arc at the spark plug. Distribution (ignition distributor) is by means of a sliding contact and bridge connection. The archetypal spark plug (two porcelain-insulated platinum wire electrodes) was invented by Lenoir. An alternative name for trembler coil, "buzzer coil," derives from the buzzing of the Ruhmkorff induction coil. Disadvantages include rapid discharge of the voltaic piles. (*See also* Figure 22 in the first chapter of this book. From Schildberger, *Bosch und die Zündung*, Stuttgart, 1952.)

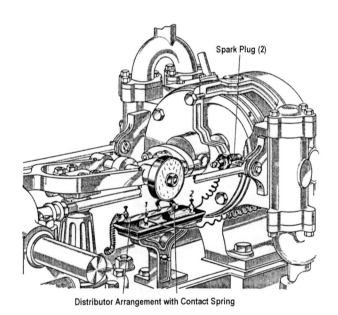

Spark Plug (2)

Distributor Arrangement with Contact Spring

Figure 226. Trembler coil ignition, Lenoir, 1860.

Flame ignition: (Otto, 1876; Figure 227) By means of appropriate settings of a sliding port, the gas/air mixture in the cylinder may be ignited by a continuously burning gas flame. The chimney ensures a steady flame. Disadvantages include problematical sealing even at relatively low pressures, and maximum speed around 200 rpm.

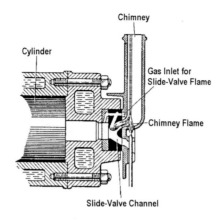

Figure 227. Flame ignition, Otto, 1876.

Low-tension magneto ignition. (Marcus, 1883; Figure 228) An armature, forced to oscillate in rotation by a crank lever, directs ignition current to a movable ignition pin in the combustion chamber, which, at the moment gas transfer takes place, is lifted away from a fixed ignition pin, forming an electrical arc between the pins. This was a predecessor to Bosch's low-tension magneto ignition of 1897 (*See* Figure 233).

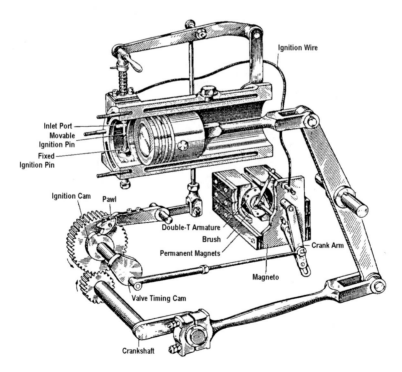

Figure 228. Low-tension magneto ignition, Marcus, 1883.

Low-tension oscillating magneto ignition. (Otto, 1884; Figure 229) A double-T armature is mounted between the pole shoes of vertically mounted bar magnets. A lever arm is briefly deflected from its neutral position by a cam. This induces a brief flow of current in the armature windings, which is directed to the ignition pin in the ignition flange. A snap-action spring, located in a barrel, quickly returns the lever arm, which moves a pushrod to break the contact between the ignition pin and ignition lever and creates a spark across the resulting gap. This was the leading ignition system for stationary engines until around 1900, able to operate independently of batteries, but permitting engine speeds of only 200 rpm.

Hot-tube ignition, untimed. (Maybach/Daimler, 1885; Figure 230) A glow tube, sealed on one end, is mounted with its open end in the combustion chamber. A burner keeps the tube at red heat. Compressed air/fuel mixture is ignited by the red-hot tube. (Timed hot-tube ignition adds a slider with ports connecting the combustion chamber and the glow tube.) Drawbacks include fire hazard, long preheat time, and difficulty starting (no ignition with a cold tube, and backfiring with an overheated tube).

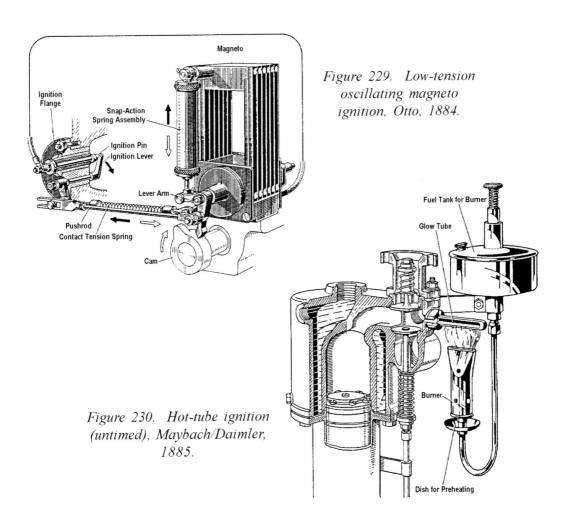

Figure 229. Low-tension oscillating magneto ignition, Otto, 1884.

Figure 230. Hot-tube ignition (untimed), Maybach/Daimler, 1885.

Trembler coil ignition. (Benz, 1885; Figure 231) In principle, similar to Lenoir's trembler coil ignition. Secondary current is controlled by an eccentric timing disc which moves a lever connected to the upper of two leaf springs. If the two leaf springs are not making contact, secondary current flows to the spark plug and creates an arc. Benz made his own spark plugs. Disadvantages are that continuously flowing primary current would drain the galvanic elements after only 10 km (6 miles).

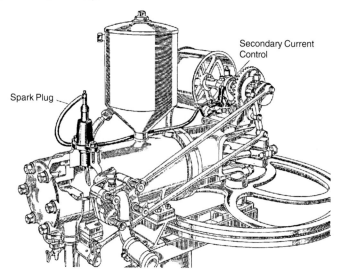

Figure 231. Trembler coil ignition, Benz, 1886.

Improved trembler coil ignition. (Benz, 1893; Figure 232) The arc is formed only when the primary circuit from the battery is closed. This occurs when contact n, mounted on timing disc m, touches contact o. Timing is adjusted manually via the pullrod q, which turns bracket p up (late timing) or down (early timing) around the axis of timing disc m.

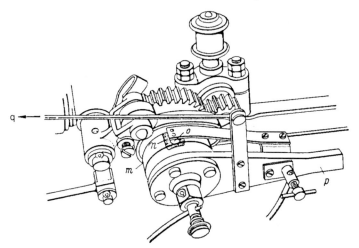

Figure 232. Improved trembler coil ignition, Benz, 1893.

Low-tension magneto ignition. (Bosch, 1897; Figure 233) Operating principle similar to that of Marcus (1883, *See* Figure 228) with forced mechanical oscillation of the armature by means of a crank arm. Installation of a light, oscillating sleeve between the more massive armature reduced oscillating mass and enabled higher engine speeds, up to approximately 1800 rpm.

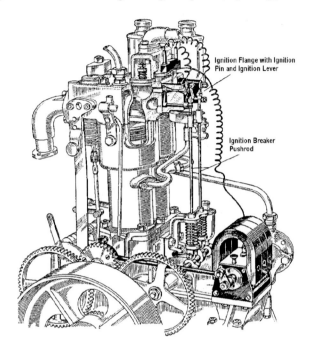

Ignition Flange with Ignition Pin and Ignition Lever

Ignition Breaker Pushrod

Figure 233. Low-tension magneto ignition, Bosch, 1897.

High-tension magneto ignition. (Bosch, 1902; Figure 234) Rotation of a double-T armature with two windings—one with few turns of heavy-gauge wire, and the other with many turns of thinner wire—produces low voltage. The heavy-gauge winding is intermittently shorted by a set of breaker points, generating high current which is then immediately interrupted again. This induces high voltage in the other winding, which jumps across the spark plug gap, making the gap conductive (ionization). As it continues turning, the same winding produces a lower voltage which flows through the now conductive spark plug gap, producing an arc. Advantages are no oscillating rods, enabling higher engine speeds, and spark plug(s) instead of make-and-break ignition pins.

Flywheel magneto ignition. (Ford, 1908–1927; Figure 235) The Ford Model T, built in the millions, had neither the high-tension magneto ignition that was customary before 1910, nor battery ignition, which entered the American scene in 1911, 15 years ahead of European practice. Instead, it had an alternating-current generator built into the flywheel. It served not only as an ignition system, but also generated electricity for vehicle lighting.

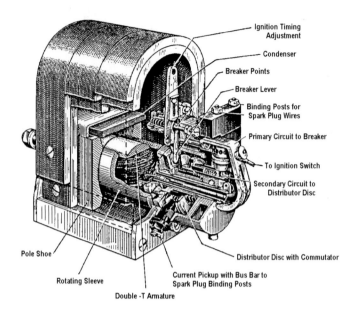

Figure 234. High-tension magneto ignition (Bosch, 1902).

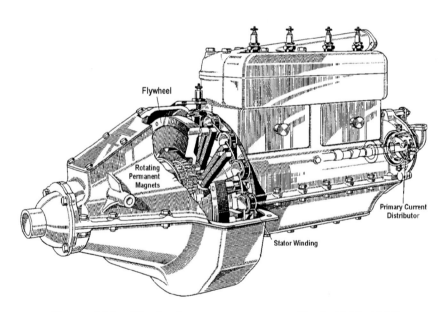

Figure 235. Flywheel magneto ignition, Ford, 1908–1927.

Battery ignition. (Bosch, 1926; Figure 236) Battery ignition consists of ignition coil, distributor, and spark plugs. Primary current flows from the battery to the coil, controlled by cam-driven breaker points in the ignition distributor. Each interruption of primary current generates a high-voltage impulse. A rotor in the distributor directs this impulse to the appropriate spark plug, where it arcs across the electrodes. In contrast to magneto ignition, the spark voltage is highest at low engine speeds and drops with rising rpm.

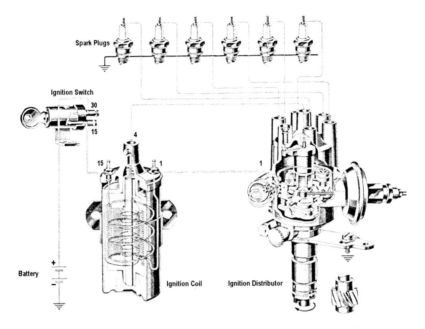

Figure 236. Battery ignition (Bosch, 1926).

Transistorized coil ignition. (TCI; Bosch, ca. 1963; Figure 237) Also known as inductive semiconductor ignition or transistorized ignition. The current normally passing through breaker points is instead handled by semiconductor components, or the points are replaced entirely by non-contacting devices. Initially, these systems still used points for switching (TI–B, transistorized ignition–breaker), later so-called "breakerless" ignition (TI–H, Hall effect, or TI–I, with inductive pulse generators).

Capacitor discharge ignition system. (CDI; Bosch, ca. 1963) In CDI ignition, also known as thyristor ignition, ignition energy is stored in a capacitor. The ignition transformer steps up the primary voltage from the capacitor discharge into the required high voltage. Well suited for rotary piston (Wankel) and high-speed, high-power reciprocating engines.

Electronic ignition. (SI, semiconductor ignition, Bosch, ca. 1974; Figure 238) Electronic or semiconductor ignition derives its name from electronic calculation of the ignition timing. An electronically stored ignition timing map replaces the familiar distributor timing curves generated

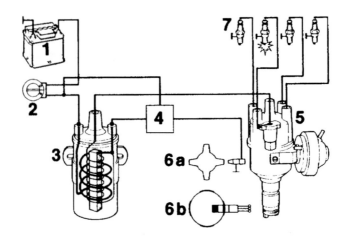

Figure 237. Breakerless TCI ignition system, Bosch, ca. 1963. 1 Battery; 2 Ignition switch; 3 Ignition coil; 4 Electronic control unit; 5 Ignition distributor with centrifugal and vacuum advance; 6(a) Inductive pickup; 6(b) Hall effect sensor (alternative); 7 Spark plugs.

by flyweights and vacuum advance or retard units. Based on the stored map, a microprocessor in the control unit processes input signals such as engine rpm and load to calculate exact ignition timing.

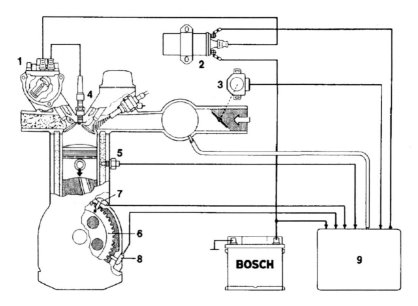

Figure 238. Electronic ignition (SI, semiconductor ignition), Bosch, ca. 1974. 1 High-tension distributor; 2 Ignition coil; 3 Throttle switch; 4 Spark plug; 5 Engine temperature sensor; 6 Flywheel; 7 Reference mark sensor; 8 Engine rpm sensor; 9 Control unit.

Distributorless electronic ignition. (BSI; Bosch, ca. 1978; Figure 239). A fully electronic ignition system is achieved by replacing the mechanical, rotating (high-tension) ignition distributor by static, electronically controlled components (e.g., dual-spark coils).

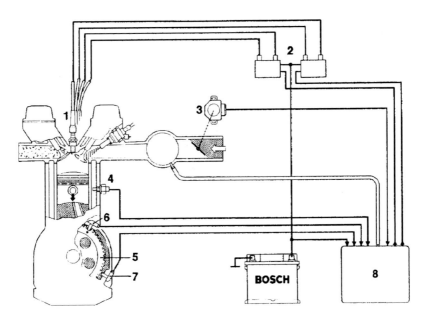

Figure 239. Distributorless, fully electronic ignition, Bosch, ca. 1978. 1 Spark plug; 2 Dual-spark coils; 3 Throttle switch; 4 Engine temperature sensor; 5 Flywheel; 6 Reference mark sensor; 7 Engine rpm sensor; 8 Control unit.

Motronic. (Bosch, 1979; Figure 240) Motronic is the result of combining electronic ignition with electronically controlled fuel injection, and represents the displacement of analog technology by digital electronics. Motronic is a further step along the road to central electronic control.

Improved trembler coil ignition. *See* Ignition systems.

Injection. *See* Diesel fuel injection; Gasoline fuel injection.

Low-tension magneto ignition. *See* Ignition systems.

Low-tension oscillating magneto ignition. *See* Ignition systems.

Lubrication. *See* Dry sump lubrication; Pressure lubrication.

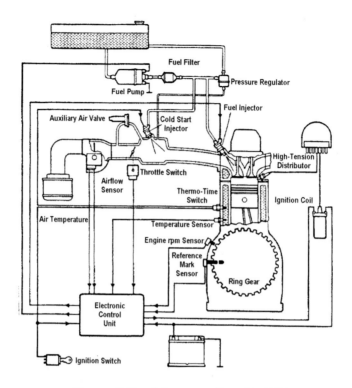

Figure 240. Motronic, Bosch, 1979.

Motronic. *See* Ignition systems.

Negative steering offset. (Figure 241) Suspension layout in which the steering axis (the imaginary line extending through the outer suspension pickup points of steered wheels) intersects the ground outboard of the wheel centerline. The purpose is to prevent sudden skidding or spinning of the vehicle in the event of uneven braking force.

On-board diagnostics (OBD). Installed on vehicles intended for the American market, OBD II continuously monitors components that affect exhaust gas composition. OBD II includes a driver display, fault code buffer, and the ability to recall fault codes by means of the repair facility's diagnostic devices. Manufacturers must certify that their emissions control systems will remain compliant for at least 100,000 miles (160,930 km). At present, OBD is not yet mandated in the European market.

Opposed-piston engine. An engine with pistons that move in opposite directions in the same cylinder, configured as two- or four-stroke, gasoline, gas, or diesel engines. Power from the upper piston is transmitted by pull rods to the lower crankshaft, or by two crankshafts joined by gearing. Made famous by Junkers (opposed-piston diesel aircraft engines), Fairbanks-Morse

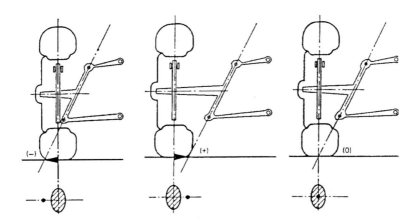

Figure 241. At left, negative steering offset, as commonly practiced in modern automobile designs; center, positive steering offset, used almost exclusively until the 1970s; right, zero steering offset (1955 Citroën and others).

(submarine and railway locomotive engines), and Napier (railway and aircraft engines). Also used as free-piston engines and gas generators.

Otto-cycle engine. Today, the term Otto-cycle engine is generally taken to mean the gasoline engine, both two- and four-stroke. As such, the term is used only to differentiate between gasoline and diesel engines.

Oversteer/understeer. In an oversteer situation, the vehicle describes a smaller radius turn (understeer: larger radius turn) than would be expected from the amount of steering lock. Oversteer is present if a tight turn is taken at excessive speed and the ratio of lateral force to axle load grows more rapidly for the rear axle than for the front axle. (Understeer: ratio of lateral force to axle load grows more rapidly for the front axle than for the rear axle.) These forces result in greater tire slip angles. (In oversteer, the rear of the car drifts outward; in understeer, the front of the car drifts outward—either of which might be a dangerous situation.) The objective of the electronic stability program (ESP) is to achieve neutral handling.

Planetary transmission. A transmission in which gears revolve around a center which is itself rotating. James Watt's steam engine employed two planetary gears. Today, planetary (also known as epicyclic) gearing is usually employed in automatic transmissions and consists of an inner "sun" gear, three "planet" gears, and an outer annular (ring) gear.

Plunger piston. In conventional reciprocating engine design, a "plunger piston" is connected directly to the crankshaft by a single connecting rod. This replaced the crosshead, familiar from steam engine practice (*See* Figure 244), with its piston, piston rod, connecting rod, and crankshaft.

Pressure lubrication. A pump draws lubricating oil from the oil sump (oilpan) and forces it through passages to individual lubricating points. From there, it flows back to the sump. Pressure lubrication replaced splash oiling around 1907 (splash oiling: the lower end of the connecting rods were fitted with buckets to transfer oil to troughs, rings, and channels, from which it flowed to lubrication points).

Radiator. *See* Honeycomb radiator; Tubular radiator.

Rotary piston engine. (Wankel engine; Figure 242) A triangular rotor mounted on an eccentric shaft rotates in a rigid, oval-shaped housing. *A, B,* and *C* indicate the three faces of the rotor, and their respective chamber volumes.

1. The intake stroke begins after the last of the spent combustion gases are forced out of chamber *A.* Compression occurs in chamber *B,* expansion in chamber *C.*

2. *A* draws in mixture, *B* compresses. In chamber *C,* burnt gases have completed their work. The exhaust port is open.

3. *A* is still drawing in mixture, *B* has reached its maximum compression. A spark ignites the compressed air/fuel mixture. Chamber *C* continues to exhaust spent gases.

4. Chamber *A* has reached its maximum volume; the intake port is closed. In chamber *B,* burning gases expand and drive the rotor around its eccentric shaft. Chamber *C* continues to exhaust spent gases.

Spray carburetor. (Maybach, 1893; Figure 243) Partial vacuum in the working cylinder created by descent of a piston (not shown) allows atmospheric air to enter at *c.* The speed of the incoming air increases at a throttle orifice *e,* and thereby draws fuel from the nozzle *f.* The float *g* and its needle valve maintain a nearly constant fuel level. Advantages over the surface carburetor include smaller size and more rapid response to changing engine speeds. The principle of Maybach's spray carburetor has been retained to the present day by all carburetor designs. (From Sass, *Geschichte des deutschen Verbrennungsmotorenbaues,* Berlin, 1962)

Steam engine. (Figure 244) A heat engine in which steam pressure drives a piston (conversion of pressure into mechanical energy). The reciprocating motion of piston, piston rod, crosshead, and connecting rod are converted into rotary motion by the crankshaft, with power taken out at the flywheel. Steam pressure is applied to the piston either on the cover side only (single-acting steam engine) or alternately on both the cover and crankshaft side (double-acting steam engine). Slide valves, poppet valves, or ports control the entry of live steam into the working cylinder and the exhaust of spent steam, either to the atmosphere (exhausting steam engine) or into a condenser (condensing steam engine). In a condenser, steam is cooled (by a tubular condenser or water injection) and returned to the boiler (e.g., reheating, vaporization, reuse of steam). Compound steam engines make use of multiple stages of expansion.

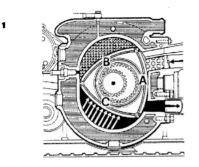

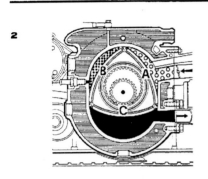

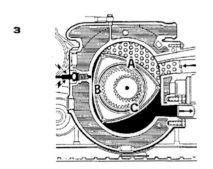

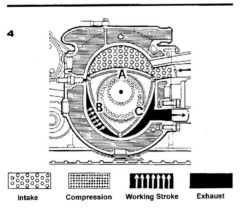

Figure 242. Wankel rotary piston engine.

Intake	Compression	Working Stroke	Exhaust

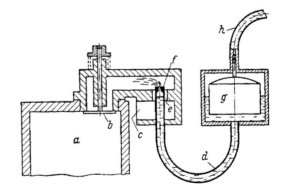

Figure 243. Maybach's spray carburetor, 1893.

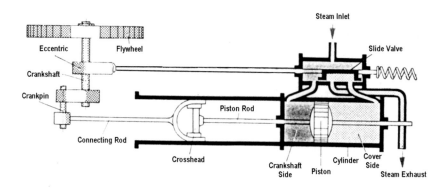

Figure 244. Schematic diagram of a double-acting steam engine.

Steering. *See* Ackermann steering; Fifth-wheel steering.

Stirling engine. *See* Air engine.

Suspension. *See* Hydropneumatic air suspension.

Thermodynamic efficiency. Thermodynamic efficiency, η, is the ratio of the useful work (or energy) produced to the energy spent. It is always less than 1 (less than 100 percent).

	ca. 1900	ca. 1970	ca. 2000
Human			
Weightlifting		←8.4–8.6→	
Walking, level surface		←33.5–35→	
Steam engine	10		
Diesel engine	26	32–40	37–45
Gasoline engine	15	22–30	32–40
Electric motor		85	90

Traffic management systems. Electronic systems have replaced traffic information reporting by police officers on the ground and in helicopters (navigation and traffic information systems of the 1970s), initially using inductive loops buried in roadways (navigation and traffic information systems of the 1980s), and since the 1990s using satellite telemetry. At present, the American Global Positioning System (GPS) is used for this purpose; as of 2008, the European Galileo satellite system may serve the same function. Along with navigation, traffic information, and traffic control, these systems may be used to direct rescue services, manage fleets, plan individual routes, collect tolls, and assist in the navigation of road and rail vehicles, marine vessels, and aircraft.

Transistorized coil ignition. *See* Ignition systems.

Transmission. *See* Direct gear; Planetary transmission.

Trembler coil ignition. *See* Ignition systems.

Tubular radiator. (Maybach, 1897; Figure 245) Hot water enters the sheet-brass radiator housing at *b*, circulates through the multitude of cooling tubes (over which air is passed), and is withdrawn at *c* by a circulating pump (water pump cooling system, as opposed to a thermosiphon system). A fan behind the radiator separates the cooling function from the motion of the vehicle and its induced airstream. The pitch of the fan blades varies from their roots outward to their tips, corresponding to their circumferential velocity. *d*. Vehicle mounting points; *e*. Filler opening; *f*. Strainer. (From Sass, *Geschichte des deutschen Verbrennungsmotorenbaues*, Berlin, 1962)

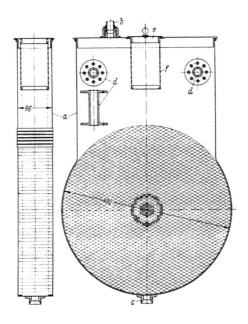

Figure 245. Maybach's tubular radiator, 1897.

Turbocharger. *See* Exhaust-driven turbocharger.

Two-stroke cycle. In two-stroke engines, the four processes that make up the thermodynamic cycle take place both above the piston (in the working cylinder) and below (crankcase).

First stroke: Compression in the cylinder, simultaneously drawing fresh air into the crankcase.

Second stroke: Work is produced in the cylinder; simultaneously, air is pre-compressed in the crankcase. Shortly thereafter, air (or air/fuel mixture) flows from the crankcase into the working cylinder and displaces spent combustion gases (exhaust).

Wankel engine. *See* Rotary piston engine.

Wheel attachment. Wheels firmly attached to the axle are called a wheel set. Wheels free to rotate on their axle are referred to as freewheeling or loose mounting. While negotiating curves, the force needed to turn a freewheeling wheel is less than that required by a wheel set. All road vehicles are freewheeling (in the sense that even powered wheels spin freely and independently of each other), whereas all rail vehicles employ wheel sets.

Worm gear drive. A screw-like gearing arrangement to reduce rotational speeds (gearing down), consisting of a worm and worm wheel. Low noise and, for rear-wheel drive, low driveline (flat driveshaft tunnel), but complex manufacturing and service.

Appendix

Industrialists and Engineers

Agnelli, Giovanni (1866–1945), Italian industrialist. Founded Fabbrica Italiana Automobili Torini (FIAT) with Carlo Biscaretti di Ruffia in 1899; first passenger car in 1899; first truck in 1903; from 1905, expanded the company into an industrial conglomerate (e.g., ball bearings, aircraft, ships, steel mills) and absorbed various other car companies.

Austin, Herbert (1866–1941), British industrialist. Manager at Wolseley Tool & Motor Car Co. in 1901; founded Austin Motor Car Co. in 1905, from 1917 Lord Austin; 1914–1918, manufactured engines and aircraft; built Austin Seven from 1922–1938 (first British mass-produced car) with licensees in Germany, France, and the United States.

Barényi, Béla (1907–1997), Austrian engineer. Studied at Wiener Technikum (Vienna Technical University); worked at Steyr, Adler, and Getefo; 1939–1974 chief of advanced projects at Daimler-Benz, where he did groundbreaking work and filed numerous patents regarding passive, pre-emptive, and active safety as applied to automotive design.

Benz, Carl (1844–1929), German engineer and industrialist. Founded a firm making a variety of technical articles in 1871; from 1877, experiments with and manufacture of two-stroke gas engines; founded Benz & Co. in Mannheim, 1883; design, construction, and rollout of the first Benz motorcar, 1885–1886; left the firm in 1903, and in 1905 founded the C. Benz & Söhne company (C. Benz & Sons) in Ladenburg.

Borgward, Carl Friedrich Wilhelm (1890–1963), German industrialist. Manufactured automobile radiators, fenders, and three-wheeled delivery cars beginning in 1919; founded Goliath Works in 1928; acquired Hansa-Lloyd works in 1929 and began manufacturing passenger cars and trucks; armaments industry, 1939–1945; began manufacturing the Hansa 1500 in 1949 (first newly designed German postwar car); manufactured the Borgward Isabella beginning in 1954 (introduction of the sporty family sedan).

Bosch, Robert (1861–1942), German industrialist. Founded a workshop for electrical components in 1886; began manufacture of magnetos and development of other ignition systems in 1887 (with Gottlob Honold); shifted from hand-crafted manufacture to industrial production in 1900; introduced the eight-hour workday in 1906; founded the Robert Bosch Hospital in 1940.

Budd, Edward Gowen (1870–1946), American industrialist. Built the first all-steel passenger car body in 1910; founded Edward G. Budd Manufacturing Company in Philadelphia, 1912; began deliveries of all-steel bodies to Oakland and Garford, 1912–1913; major orders from Dodge and others beginning in 1914; in 1926–1928, acquired the German coachbuilder Ambi (Berlin), Lindner (Ammendorf), and DIW (Berlin).

Büssing, Heinrich (1843–1929), German industrialist. After studying mechanical engineering as an "auditor," built and sold velocipedes (1868–1870), and railway signaling systems (1871–1913); founded Germany's first factory dedicated to production of motor trucks and buses in 1903; established a bus line and transit company in 1904; from 1905, granted licenses to firms in England, Bavaria, Austria, and others; in 1914, issued design contract to independent artist.

Chrysler, Walter P. (1875–1940), American engineer and industrialist. Founded Chrysler Corp. in 1920 after working for various railroads, General Motors, and Willys Overland; first Chrysler automobile in 1924; acquired Dodge in 1928, and introduced two new marques (Plymouth and De Soto); introduced a streamlined car in 1934 (Airflow).

Citroën, André Gustave (1878–1935), French industrialist. Established gear and transmission plant in 1902; acquired Automobiles Mors in 1908; built artillery shell factory in 1915; introduced assembly line production to Europe in 1919 (Citroën Type A); introduced small car (5 CV) and crossed the Sahara using halftracked vehicles in 1921; crossed equatorial Africa (Croisière Noire) in 1924–1925, and Asia (Croisière Jaune) in 1931–1932; founded Citroën Köln (Cologne, Germany) in 1928; produced the first diesel passenger car in 1934.

Cugnot, Nicolas Joseph (1725–1804), French engineer. Trained in Germany, studied the mechanical engineering texts of Jakob Leupold; in Paris, published papers on fortress construction and military engineering; in 1769, built an experimental steam car; completed a load-carrying steam car (*fardier*) for military application in 1771.

Cummins, Clessie (1888–1968), American industrialist. Founded Cummins Engine Co. in 1919; built diesel engines for marine applications; built prototype diesel passenger car and link-belt tractor in 1930; built diesel record-setting and race cars, 1930–1934; produced diesel engines for commercial vehicles; invented retarder brake using shifted exhaust valve timing (Jacobs brake), in production since 1961.

Daimler, Gottlieb Wilhelm (1834–1900), German engineer and industrialist. Trained in England and Germany, 1861–1872; technical director of the Deutz Gas Engine Works, 1872–1882; independent experimental workshop in Cannstatt (Stuttgart), 1882–1889; design (with Wilhelm Maybach) and construction of light gasoline engines (two-wheeler in 1885, motor carriage in 1887); founded Daimler Motoren Gesellschaft (Daimler Engine Company)—DMG—in 1890.

Daimler, Paul (1869–1945), German engineer. Son of Gottlieb Daimler; technical director of DMG from 1897–1902; technical director of Österreichische Daimler-Motoren KG (Austrian Daimler Engine Co., later Austro-Daimler) from 1902–1905; designed motor vehicles and engines for airships, airplanes, and submarines from 1905–1922; 1922–1928, technical director of Horch; 1929–1945, consulting engineer.

de Dion, Count Albert (1856–1946), French industrialist. Built steam boilers (with Georges Bouton and Charles Trépardoux), 1882; built steam three-wheeler, 1883; patented the de Dion-Bouton rear axle, 1893; founded De Dion & Cie, 1894; built first gasoline engine in 1895, available for custom installation; built tri- and quadricycles from 1896; built the world's first V8-powered production car, 1910–1923; ceased passenger car production in 1932, commercial vehicle production in 1950.

Diesel, Rudolf (1858–1913), German engineer. Studied at Munich Polytechnic from 1875–1880; in 1881, director for Linde's Ice Machines in Paris and designed an ammonia engine; 1890, director of Linde's in Berlin and began work on his own combustion engine (later known as the diesel engine); granted patent number 67 207 for diesel engine in 1892; began construction of experimental diesel engines in 1893.

Dunlop, John Boyd (1840–1921), British veterinarian and inventor. Re-invented the pneumatic tire in 1888 (initially for bicycles); founded Dunlop Rubber Co. in London, 1889; established Dunlop Gummi Co. in Hanau (near Frankfurt, Germany) in 1893; produced wire-bead pneumatic tires for automobiles in 1894.

Durant, William Crapo (1862–1947), American industrialist. Founded General Motors in 1908; Chevrolet in 1911; combined several auto firms to form Durant Motors (1921–1932).

Duryea, Charles E. (1861–1939) and **J. Frank** (1869–1967), American mechanics and industrialists. The Duryea brothers built the first working gasoline-powered American automobile in 1893; founded Duryea Motor Wagon Co. and won the *Chicago Times-Herald* contest in 1895; the brothers parted professionally in 1900. Charles continued with the firm; in 1902, granted license to Britain; in 1908–1913, built a high-wheeler (Buggyaut); ended production in 1917. Frank, working at Stevens-Duryea, built two-cylinder, later four-cylinder (1905), and six-cylinder (1906) passenger cars in the high-price classes; production ended in 1927.

Earl, Harley J. (1893–1969), American designer. In 1919, built Cadillac custom bodywork for a local dealer catering to Hollywood film stars; hired by General Motors in 1926 for the sole purpose of designing a smaller Cadillac, sold as the LaSalle in 1927; success of the LaSalle prompted General Motors to establish a dedicated styling department, an industry first, with Earl as director (General Motors Art & Colour Section; Chrysler established its own styling studio in 1932, Ford in 1935, and Studebaker in 1936); Earl introduced clay as the medium for three-dimensional prototypes to supplement two-dimensional renderings, as well as the General Motors Motoramas, which presented Earl-designed prototypes ("Dream Cars") and production vehicles within a singing, dancing tableau (Figure 140); worked at General Motors until 1959.

Ford, Henry (1863–1947), American engineer and industrialist. First experimental car, 1896; founded Ford Motor Co. in 1903; built Model T (Figure 50) from 1908–1927, with 15 million examples sold, the most successful car in history until 1972; introduced assembly line production in 1913; tractor in 1916; truck based on Model T in 1917; first mass-produced car with a V8 engine in 1932 (Figure 93); established the Ford Foundation in 1936 to promote education and vocational training.

Horch, August (1868–1951), German engineer and industrialist. Plant manager at Benz & Cie, 1896–1899; founded Horch Motorwagenwerke in 1899; founded Audi Werke in 1909; named honorary doctor of engineering in 1922; authored *Ich Baute Autos* ("I built cars") in 1937; member of the Auto Union management board in 1942.

Imbert, Georg(es) (1884–1950), Lorrainer chemist. Salaried and self-employed chemist in Nuremberg, Alsace, and Manchester 1905–1922; gas generator designer at de Dietrich 1923–1926; in 1927, founded his own gas generator firm in Alsace; granted licenses to Imbert GmbH of Cologne from 1931–1934.

Issigonis, "Sir Alec" Alexander Arnold Constantine (1906–1988), Turkish-British engineer. Draftsman at Humber, 1934–1936; Morris, 1936–1952, where he designed the Minor in 1948 (Figure 144); Alvis, 1952–1956, where he designed luxury cars; BMC/BLMC, 1956–1971, where he designed the Mini in 1959 (Figure 156); knighted in 1969 (Sir Alec Issigonis).

Jaray, Paul (1889–1974), Austrian engineer. Beginning in 1912, designed seaplanes and airships; in 1920, scientific investigations into road vehicle aerodynamics; first patents in 1921; from 1922, streamlined cars built to Jaray patents; Jaray patents evaluated in the United States, 1931–1938, by the Jaray Streamline Corp. of New York, and in Europe, 1933, by the AG für Verkehrspatente (Corporation for Vehicle Patents), Lucerne, Switzerland.

Jenatzy, Camille (1868–1913), Belgian designer and racer. From 1898, electric and gasoline-electric hybrid cars; in 1899, record car *La Jamais Contente* ("Never Satisfied," Figure 58); 1903, victory in the Gordon Bennett Cup in Ireland aboard Mercedes 60 HP (Figure 47); additional sports successes.

Lankensperger, Georg (1779–1847), Bavarian wagonmaker and inventor. Began building vehicles in Munich, 1806; from 1811, built vehicles and sleighs for the Bavarian court; in 1816, invented what is now known as Ackermann steering; in 1839, founded a pilgrimage rectory; in 1845, founded a school for young girls.

Ledwinka, Hans (1878–1967), Austrian designer. 1897–1902, designer at the Nesselsdorfer Wagenbau-Fabriks-Gesellschaft (Nesselsdorfer Wagonbuilding Works Inc., later Tatra); 1902–1905, steam car design; in 1905, technical director of Tatra; 1917–1921, chief engineer of Steyr; 1921–1945, technical director of Tatra (passenger cars and commercial vehicles; central tube frame, air-cooled engines, and streamlining).

Lenoir, Jean Joseph Etienne (1822–1900), self-taught Luxembourgian inventor. Invented enameling technology, electrotyping, telecopying, and electric railway brakes and signals; built the first functional gas engine, 1860 (Figure 22); installed such an engine in a boat, 1861–1865, and in a break in 1863 (Figure 23).

Loewy, Raymond (1893–1986), Franco-American industrial designer. Trained in Paris; New York-based designer and consultant from 1919; furniture, household appliances, food packaging machinery (Flexvac), commercial packaging (Coca-Cola bottle shape), wheeled and tracked tractors, express train locomotives (Pennsylvania Railroad S1, 1937), Lucky Strike cigarette packaging (1940), warehouses, supermarkets, and train, ship, and aircraft interiors, to name only a few. Beginning in 1939, designs for Studebaker, most notably the classic bullet-nosed 1950 Champion/Commander (Figure 139), 1953 "European Look" Starliner coupe, and the 1963 Avanti.

Luthe, Claus (*1932), German body and auto designer. 1948–1955, training, master's and engineer's papers; 1955–1956, body engineer and designer at Fiat/Heilbronn; 1956–1971, designer and chief designer at NSU (Ro 80, Figure 163); 1971–1976, director of the Audi design department; 1976–1990, chief designer at BMW.

Marcus, Siegfried (1831–1898), German-Austrian mechanic and inventor. From 1860, numerous inventions, including the magneto blasting machine (variously known as the "Viennese Igniter," plunger, or rack-bar exploder) in 1864; design of a gasoline engine for hand carts in 1870; brush carburetor with preheat in 1882; magneto ignition in 1883; and a second Marcus car in 1888.

Maybach, Wilhelm (1846–1929), German designer. 1861–1872, education and training including self-study and work as a mechanical design engineer; from 1869, co-worker with Gottlieb Daimler; 1872–1882, chief designer at Gasmotorenfabrik Deutz; 1882–1895, vehicle engines in Daimler's experimental workshop, as well as outside the Daimler Motoren Gesellschaft (DMG); 1895–1907, technical director of DMG; in 1909, established Luftfahrzeug-Motoren GmbH (Aircraft Engines Ltd.) at Bissingen, with son Karl and Graf Zeppelin.

Morris, William Richard (1877–1963), British industrialist and philanthropist. Built bicycles in 1893; first car in 1913 ("Bullnose" Oxford); in the 1920s, instituted price reductions, captured 50 percent share of British market, and takeover of MG, Hotchkiss of Britain, Léon Bollée of France, Wolseley, and Riley (1938); created the first Viscount Nuffield in 1938; benefactor; built Minor (1949, Figure 144) and Mini (1959, Figure 156).

Olds, Ransom Eli (1864–1950), American industrialist. Beginning in 1891, built one-off steam, gasoline, and electric cars; founded Olds Motor Vehicle Co. in 1897; 1900–1904, Curved Dash Oldsmobile (first mass-produced car, Figure 49); founded Reo Motor Car Co. in 1904; Reo commercial vehicles, 1908–1967; affordable Wolverine marque produced by Reo, 1927–1929; Reo passenger car production ended in 1936.

Ostwald, Fritz (1913–1999), German physicist. Developed non-locking, non-intermittent brake controller in 1940 (predecessor of ABS); 1950–1978, leading positions in the fields of hydraulics, steering, research, and advanced projects at Teves; in 1958, developed negative steering offset and five-link suspension.

Otto, Nikolaus August (1832–1891), German businessman and inventor. 1860–1862, further development of Lenoir's gas engine; 1862–1867, design and construction of atmospheric gas engines; in 1864, with Eugen Langen, founded N.A. Otto & Cie, as of 1872 Gasmotoren-Fabrik Deutz, later Klöckner-Humboldt-Deutz AG; invented the four-stroke engine in 1876; magneto make-and-break ignition in 1884.

Papin, Denis (1647–1712 or 1714), French natural philosopher. Gunpowder engine in 1688 and atmospheric steam engine in 1690, both built in Hesse; in 1681, invented the steam pressure cooker, with pressure-tight lid and safety valve; in 1706, invented direct-acting steam pump.

Pininfarina (Giovanni Battista Farina, 1893–1966), Italian auto designer. Initially worked in his brother's coachbuilding firm, Stabilimenti Farina; from 1930, Carrozzeria Pininfarina; until 1940, coachbuilder for Lancia, Fiat, Alfa Romeo, Isotta-Fraschini, and Hispano-Suiza; after World War II, 1952 Nash Golden Airflyte, Ferrari (from 1952), Peugeot (from 1955), 1959 Fiat 1800 (Figure 141), BMC, General Motors, and others, as well as transition to prototype and mass production; was allowed to change his name from Farina to Pininfarina by the Italian government in 1961; established research and development center in 1966.

Porsche, Ferdinand (1875–1951), Austrian mechanical designer and entrepreneur. Lohner-Porsche electric car, 1900; Austrian Daimler Motors, 1906–1923; Daimler Motoren Gesellschaft/Daimler-Benz AG, Stuttgart, 1923–1929; Steyr Works, 1929–1930; 1931–1944, Porsche independent design bureau in Stuttgart (1934, Auto Union race car; 1935, KdF (Volkswagen and others); in 1947, design and manufacture of sports cars in Gmünd, Austria; from 1950 in Stuttgart.

Probst, Karl Knight (1883–1963), American engineer. After completing studies and a tenure in France, worked for Chalmers, Lozier, Peerless, Milburn Electric, Reo, and various industry suppliers; from 1935, independent engineering office; in five days (July 18–22, 1940), designed a General Purpose Vehicle (GPV, Jeep) for military applications under contract to American Bantam and built 70 prototypes within 49 days; from 1941, designs for Chrysler, Kaiser, Ford, and Reo.

Renault, Louis (1877–1944), French mechanic and industrialist. First Renault car, based on de Dion-Bouton, 1898; founded Renault Frères, 1899; from 1900, participated in road races and expanded firm to become one of the largest French auto makers.

Ricardo, Sir Harry Ralph (1885–1974), British engineer. Worked in an engineering office, 1907; improved the gasoline engine used in British tanks; established independent design office in 1919; carried out research into combustion processes and lubricants, and developed the Comet Turbulent Cylinder Head and diesel engine; knighted in 1948.

Rumpler, Edmund (1872–1940), Austrian mechanical designer and industrialist. Until 1907, designer with various auto makers (e.g., Nesselsdorfer, Daimler-Marienfelde, and Adler); founded Rumpler Luftfahrzeuge GmbH (Rumpler Aircraft Ltd.) in 1908 (built Rumpler Taube and others); from 1921, Rumpler Tropfenwagen, front-drive passenger cars and commercial vehicles, countless patents for motor vehicles and aircraft.

Vollmer, Joseph (1871–1955), German mechanical designer. Worked for Bergmann Industriewerke Gaggenau (Orient-Express), 1894–1898; 1898–1902, Kühlstein-Wagenbau (electric and gasoline cars, trucks, attachments); 1902–1906, with NAG (passenger car and DURCH road train, Figure 67); 1906–1936, independent automotive design company (gasoline and diesel engines, 1916 A7V tank, 1917–1918 light armored cars, 1919 link-belt tractors for Hanomag, Podeus, and Dinos).

von Guericke, Otto (1602–1686), German natural philosopher and statesman. Councilman (1626); engineer (1631); representative at peace negotiations; built his "Magdeburg spheres" in 1656 to demonstrate the power of air pressure; in 1661, demonstrated the potential of a piston to do useful work when driven into an evacuated cylinder by air pressure.

von Herkomer, Sir Hubert (1849–1914), German-English artist. Studied in London and Munich; portrait painter to the British royal family, the aristocracy, and the landed gentry; knighted in Bavaria and Britain; countless awards; established the 1905–1907 "Herkomer Trials" to advance the state of touring cars.

Wankel, Felix (1902–1988), German self-taught engineer and inventor. From 1924, involved with rotary piston machines; in 1936, established Wankel experimental laboratory in Lindau, developed control systems and sealing components; in 1951, cooperative effort with NSU; in 1957, first experimental Wankel rotary piston engine (NSU); in 1964, first production car with Wankel engine (NSU Spider, Figure 170).

Watt, James (1736–1819), Scottish mechanic and inventor. 1763–1764, experimented with Newcomen steam engine; was granted British patent number 913 for a single-acting steam engine in 1769; first Boulton & Watt steam engine placed in service in 1774; issued patent for double-acting steam engine in 1782; patent for parallelogram linkage in 1784.

Winton, Alexander (1860–1932), Scottish-American industrialist. In 1890, bicycle manufacturing; in 1896, first motorcar; in 1897, founded Winton Motor Carriage Co.; from 1900, single-, four-, and eight-cylinder race cars; in 1903, first transcontinental run aboard a Winton; from 1904, four- and six-cylinder production cars; from 1913, diesel engines for ships, freighters, generators, and railroads; acquired by General Motors in 1923.

Bibliography

Benz, Carl, *Lebensfahrt eines Erfinders*, Leipzig, 1925.

Biscaretti di Rufia, Rodolfo, *Il Museo dell'Automobile Torino*, Turin, 1966.

Bröhl, H.P., *Paul Jaray, Stromlinienpionier*, Bern, 1978.

Caunter, C.F., *The Light Car*, London, 1970.

Conservatoire National des Arts et Métiers. *Transports sur Route*, Paris, 1953.

Conservatoire National des Arts et Métiers. *La Voiture à Vapeur de Cugnot*, Paris, 1956.

Cummins, C. Lyle, Jr., *The Diesel Odyssey of Clessie Cummins*, Wilsonville, Oregon, 1998.

Cummins, C. Lyle, Jr., *Diesel's Engine*, Wilsonville, Oregon, 1993.

Daimler-Benz AG, *75 Jahre Nutzfahrzeug-Entwicklung 1896–1971*, Stuttgart, 1971.

Daimler Motoren Gesellschaft, *Zum 25jährigen Bestehen der DMG*, Stuttgart, 1915.

Davison, C.St.C.B., *History of Steam Road Vehicles*, London (Science Museum), 1970.

Deutscher Automobil Club, *Gordon Bennett Führer 1904*, Munich, 1904.

Diesel, Eugen, *Die Geschichte des Diesel-Personenwagens*, Stuttgart, 1955.

Diesel, Eugen; Goldbeck, Gustav; and Schildberger, Friedrich, *From Engines to Autos: Five Pioneers in Engine Development and Their Contributions to the Automotive Industry*, Chicago, 1960.

Eckermann, Erik, *Die Achsschenkellenkung und andere Fahrzeug-Lenksysteme*, Munich, 1998.

Eckermann, Erik, *Automobile.Technikgeschichte im Deutschen Museum*, Munich, 1989.

Fersen, Hans-Heinrich von, *Autos in Deutschland 1885–1920*, Stuttgart, 1965.

Fersen, Olaf von, *Ein Jahrhundert Automobiltechnik—Personenwagen*, Düsseldorf, 1986.

Ford, Henry, with Crowther, Samuel, *My Life and Work*, New York, 1922.

Georgano, G.N., *The Complete Encyclopaedia of Motorcars 1885–1968*, New York, 1968.

Georgano, G.N., *The Complete Encyclopedia of Commercial Vehicles*, Osceola, Wisconsin, 1978.

Ginzrot, Johann Christian, *Die Wägen und Fahrwerke der Griechen und Römer und anderer alter Völker*, Vols. I and II, Munich, 1817.

Glaser, Hermann, *Das Automobil. Eine Kulturgeschichte in Bildern*, Munich, 1986.

Goebel, G., *Automobil-Motoren*, Vienna, 1905.

Horch, August, *Ich baute Autos*, Berlin, 1937.

Kimes, Beverly Rae, *The Star and the Laurel.* Montvale, New Jersey, 1986.

Kirchberg, Peter, *PS-Veteranen*, Berlin, 1965.

Klemm, F.; Roosen, R.; Treue, W., *200 Jahre Industrielle Revolution*, Munich, 1969.

Klemm, Friedrich, *Die alte Technik in Bilddokumenten*, Munich, 1977.

Klemm, Friedrich, *A History of Western Technology*, London, 1959.

Klemm, Friedrich, *Kurze Geschichte der Technik*, Freiburg, 1961.

Knust, Dieter, *The Karmann Story: Germany's Coachbuilder to the World*, Osnabrück, 1996.

Koenig-Fachsenfeld, Reinhard, *Aerodynamik des Kraftfahrzeugs*, Frankfurt/M., 1951.

Koeppen, Hans, *Im Auto um die Welt*, Berlin, 1908.

Kühner, Kurt, *Geschichtliches zum Fahrzeugantrieb*, Friedrichshafen, 1965.

Lehr, Wilhelm, *Kleine Chronik der Motorenzündung*, Stuttgart, 1966.

Lichtenstein, C.; and Engler, F., editor, *Stromlinienform*, Stuttgart, 1992.

Lichtenstein, Claude, and Engler, Franz, editor, *Streamlined: The Aesthetics of Minimized Drag. A Metaphor for Progress*, Baden, Switzerland, 1995.

Loewy, Raymond, *Never Leave Well Enough Alone*, New York, 1951.

Mayer-Larsen, Werner, *Der Untergang des Unternehmers*, Munich, 1978.

Nader, Ralph, *Unsafe at any Speed: The Designed-In Dangers of the American Automobile*, New York, 1965.

Nelson, Walter Henry, *Small Wonder: The Amazing Story of the Volkswagen*, Boston, 1970.

Oertel, Walter, *Der Motor in Kriegsdiensten*, Leipzig, 1906.

Oppel, Frank, editor, *Motoring in America: The Early Years*, Secaucus, New Jersey, 1989.

Ostwald, Walter, *Autler-Chemie*, Berlin, 1910.

Ostwald, Walter, *Generator-Jahrbuch*, Berlin, 1942.

Oswald, Werner, *Kraftfahrzeuge und Panzer der Reichswehr, Wehrmacht und Bundeswehr*, Stuttgart, 1973.

Otzen, Robert, *Die Autostrasse Hamburg-Frankfurt-Basel*, Hannover, ca. 1927.

Pearson, Henry Clemens, *Rubber Tires and All About Them; Pneumatic, Solid, Cushion, Combination, for Automobiles, Omnibuses, Cycles, and Vehicles of Every Description*, New York, 1906.

Rae, John B., *The American Automobile*, Chicago, 1965.

Reichardt, Hans, *Berliner Omnibusse*, Düsseldorf, 1975.

Reismann, Otto, *Reichsautobahnen*, Frankfurt/M., 1935.

Ricardo, Harry R., *The High-Speed Internal-Combustion Engine*, London, 1944.

Ricardo, Harry R., *Schnellaufende Verbrennungsmaschinen*, Berlin, 1926.

Richard, Yves, *Renault 1898–1965*, Paris, 1965.

Robert, Jean, *Histoire des Transports dans le Ville de France*, Paris, 1974.

Sass, Friedrich, *Geschichte des deutschen Verbrennungsmotorenbaues*, Berlin, 1962.

Schildberger, Friedrich, *Bosch und die Zündung*, Stuttgart, 1952.

Schildberger, Friedrich, *Daimler und Benz auf der Pariser Weltausstellung 1889*, Stuttgart, 1964.

Schildberger, Friedrich, *Entwicklungsrichtungen der Daimler- und Benz-Arbeit bis um die Jahrhundertwende*, Stuttgart, ca. 1962.

Schildberger, Friedrich, *Gottlieb Daimler, Wilhelm Maybach und Karl Benz*, Stuttgart, 1968.

Schmidt, Ernest, *Automobiles Suisses*, Chateau de Grandson, 1967.

Schuster, George, with Mahoney, Tom, *The Longest Auto Race: New York to Paris, 1908*, New York, 1966.

Science Museum, *Steam Road Vehicles*, London, 1953.

Sedgwick, Michael, *Cars of the 1930s*, London, 1970.

Sedgwick, Michael, *FIAT*, London, 1974.

Seherr-Thoss, H.C. Graf von, *Die Deutsche Automobilindustrie*, Stuttgart, 1974.

Seper, Hans, *Damals als die Pferde scheuten*, Vienna, 1968.

Seper, Hans, *Siegfried Marcus und seine Verbrennungsmotoren*, Vienna, 1974.

Spielberger, Walter, *Hanomag Sd.Kfz 251/1 APC*, Surrey, 1967.

Treue, Wilhelm, editor, *Achse, Rad und Wagen*, Göttingen, 1986.

Vanderveen, Bart H., *M 3 Half Track APC*, Armour in Profile, Surrey oJ.

Vanderveen, Bart H., *Tanks and Transport Vehicles World War II*, London/New York, 1974.

Vitra Design Museum, *Automobility*, Weil am Rhein, 1999.

Wankel, Felix, *Einteilung der Rotationskolbenmaschinen*, Stuttgart, 1963.

Weymann, Werner, *Untersuchungen des Gaswechsels und der Leistung an Historischen Daimler-Motoren aus den Jahren 1885/86 und 1889*, Stuttgart, 1961.

Wherry, Joseph H., *Automobiles of the World*, New York, 1968.

Wise, David Burgess, *The Motor Car*, London and New York, 1977.

Zatsch, Angela, *Staatsmacht und Motorisierung am Morgen des Automobilzeitalters*, Konstanz, 1993.

Zeller, Reimar, *Automobil. Das magische Objekt in der Kunst*, Frankfurt/M., 1985.

Yearbooks, Handbooks, and Lexica

ADAC, *Technisches ADAC-Jahrbuch 1933–34*, Munich, 1934.

Automobile Quarterly, *The American Car Since 1775*, New York, 1971.

Automobiltechnischer Kalender/Automobiltechnisches Handbuch 1907–44, Berlin, various editions.

Bosch, Robert GmbH, *Automotive Handbook*, Warrendale, Pennsylvania, 1996.

Bosch, Robert GmbH, *Kraftfahrtechnisches Taschenbuch*, Stuttgart, various editions.

Brockhaus' Konversations-Lexikon, 16 Bände, Leipzig, 1892.

Buschmann, H., *Taschenbuch für den Auto-Ingenieur*, Stuttgart, 1942.

Buschmann, H., and Koessler, P., *Handbuch für den Kraftfahrzeugingenieur*, Stuttgart, 1963 and 1973.

DDAC, *Das Technische DDAC Jahrbuch 1938*, Frankfurt/M., 1938.

Meyer, *Meyers Enzyklopädisches Lexikon*, 25 volumes, 1 Vol. Addenda, Mannheim, 1971.

RDA/VDA, *Tatsachen und Zahlen*, Berlin, Frankfurt/M., various years.

Periodicals

ADAC Motorwelt, Munich.

Allgemeine Automobil-Zeitung, Berlin.

Allgemeine Automobil-Zeitung, Vienna.

Automobil- und Motorrad-Chronik, Munich.

Automobil Revue, Bern.

Automobil Revue Katalog, Bern.

Automobile Quarterly, New York.

Automobilwelt und Flugwelt, Berlin.

Autotechnische Zeitschrift ATZ, Stuttgart.

Cars & Parts, Sidney, Ohio.

Classic & Sports Cars, London.

Deutsche Fahrzeug-Technik, Gera.

Das Fahrzeug, Eisenach.

Der Motorfahrer, Munich.

Der Motorwagen, Berlin.

Motor-Jahr, Berlin (East).

Motor-Kritik, Darmstadt.

Motor Rundschau, Frankfurt.

Motor Schau, Berlin.

Motortechnische Zeitschrift MTZ, Stuttgart.

Neue Kraftfahrer Zeitung, Stuttgart.

Old Motor, London.

Special Interest Autos, Stockton, California.

Spiegel, Hamburg.

Süddeutsche Zeitung, Munich.

Thoroughbred & Classic Cars, London.

Veteran & Vintage Magazine, London.

Diverse corporate magazines.

Author's personal collection.

Figure Credits

Figure Number(s)

Audi

166, 170, 173, and 213

Automobil-Revue

154, 159, 164, 167, and 208

Bartels

183

BMW

171

Chrysler

106 and 218

Citroën

99, 116, 134, 145, and 155

Cummins

128

Daimler-Benz/Chrysler

30, 33, 34, 46, 47, 62, 64, 119, 123, 126, 149, 150, 160, 192, 198, and 203

Deutsches Museum

1–4, 8, 11, 14, 17, 20, 24, 28, 31, 32, 35–38, 40, 41, 43, 48, 50, 54, 58-61, 65-68, 70, 76, 81, 82, 90–92, 95, 97, 98, 102, 105, 109, 111–115, 122, 129, 186, and 204

Eckermann archives

From Ginzroth: D*ie Wägen und Fahrwerke der Griechen und Römer*: 5–7, 9, and 10

From Sass: *Geschichte des deutschen Verbrennungsmotorenbaues*: 15, 22, 26, 27, 29, 222, 224, 225, 228, 229, 232, 243, and 245

From *Bosch und die Zündung*: 21, 226–231, and 233–236

Eckermann archives *(cont.)*

From *Bosch Automotive Handbook:* 237–240

12, 13, 16, 18, 19, 23, 25, 39, 42, 44, 45, 49, 51–53,
55–57, 69, 71–75, 77–80, 83–89, 94, 96, 100, 101,
103, 104, 107, 108, 110, 117, 121, 124, 130–133, 135–
140, 144, 146–148, 152, 153, 156, 162, 163, 175, 178,
182, 184, 185, 187, 195, 206, 211, 214, 217, 219, 220,
241, 242, and 244

FFG Falkenried   188

Fiat   141 and 181

Ford   93, 118, and 193

General Motors   157 and 161

Hanomag IG   120 and 127

Honda   169, 207, and 209

Kässbohrer   189

Kia   176

Kittler   143

Lada   179

lastauto omnibus   223

Dr. Hans Mai   63

Mazda   201

Mitsubishi   168 and 174

Motorpresse   191

Neoplan   190 and 205

Opel   196 and 199

PSA	202 and 216
Porsche	215
Raffay	194
Ricardo	125
Saab	172
Siemens	212
Smart	221
Suhr	200
Volkswagen	142, 151, 158, 165, 177, 180, and 210
Winter	197

Abbreviations

AAZ	Allgemeine Automobil-Zeitung/General Automobile News
ABOAG	Allgemeine Berliner Omnibus-Aktien-Gesellschaft, Berlin/General Omnibus Joint-Stock Company, Berlin 1
AERA	American Electric Railway Association
ALAM	Association of Licensed Automobile Manufacturers
ASF	Audi Space Frame
ATZ	Automobiltechnische Zeitschrift/Journal for Automotive Technology
A U	Abgasuntersuchung/Exhaust Emissions Test
BVG	Berliner Verkehrs-Aktien-Gesellschaft/Berlin Transit Corporation
CAFE	Corporate Average Fuel Economy
CAN-bus	Controller Area Network, i.e., Multiplex wiring
CARB	California Air Resources Board
ckd	Completely knocked down
COMECON	Council for Mutual Economic Assistance
DAF	Deutsche Arbeitsfront/German Workers' Front
DAT	Deutsche Automobil Treuhand/German Automobile Trust
DDAC	Der Deutsche Automobil-Club/The German Automobile Club

dohc	Double overhead camshaft
DOT	Department of Transportation

EEC	European Economic Community
EFTA	European Free Trade Association
EPA	Environmental Protection Agency
ESV	Experimental Safety Vehicle
EU	European Union
Euro-NCAP	New Car Assessment Program

FEM	Finite Element Methods
FKFS	Forschungsinstitut für Kraftfahrwesen und Fahrzeugmotoren an der Technischen Hochschule Stuttgart/Research Institute for Motor Vehicles and Vehicle Engines at the Technical University of Stuttgart, Germany

GATT	General Agreement on Tariffs and Trade

HAFRABA	Verein zur Vorbereitung der Autostrasse Hamburg-Frankfurt-Basel/Organization to Prepare the Autobahn Hamburg-Frankfurt-Basel
HMMWV	High Mobility Multipurpose Wheeled Vehicle (Hummer)
HWA	Heereswaffenamt/Army Ordnance Office of the Reich

IAA	Internationale Automobil-Ausstellung/International Automobile Exhibition, Frankfurt, Germany
IAMA	Internationale Automobil- und Motorrad-Ausstellung/International Automobile and Motorcycle Exhibition

KdF	Kraft Durch Freude/Strength Through Joy (Nazi organization)
LEV	Low Emissions Vehicle
LGOC	London General Omnibus Company
MIT	Massachusetts Institute of Technology
NSDAP	Nationalsozialistische Deutsche Arbeiterpartei/National Socialist German Workers' Party
NSKK	Nationalsozialistisches Kraftfahr-Korps/National Socialist Motorist Corps
OBD	On-Board Diagnostic
ohc	Overhead camshaft
OHL	Oberste Heeresleitung/German High Command (World War I)
ohv	Overhead valve engine
OPEC	Organization of Petroleum Exporting Countries
QMC	Quartermaster Corps
RDA	Reichsverband der Automobilindustrie/Imperial Automobile Industry Association (Germany)
rpm	(Crankshaft) Revolutions per minute
SA	Sturmabteilung/(Nazi) Storm troopers
SS	Schutzstaffel/(Nazi) Home-defense squadron

StVZO   <u>St</u>rassen<u>v</u>erkehrs-<u>Z</u>ulassungs-<u>O</u>rdnung/Motor Vehicle Code (Germany)

SUV   Sport Utility Vehicle

sv   Side-valve engine

VÖV   Verband Öffentlicher Verkehrsbetriebe/Association of Public Transit Systems

ZEV   Zero Emissions Vehicle

Index of Names

Persons

Index of Names

Brands, Manufacturers, Types, and Organizations

Index of Names
Sporting Events and Sites

Keywords Index

About the Author

Erik Eckermann was born in Hamburg in 1937. After high school, he trained as an automotive mechanic and export clerk. From 1959 to 1963, Mr. Eckermann was employed first as a spare parts salesman and later as manager of a factory-authorized Simca/Auto Union/Krupp workshop in Ghana, West Africa. From 1963 to 1965, he held posts in technical capacities at Daimler-Benz dealerships in Duisburg and Berlin, Germany.

Mr. Eckermann took up studies at Hamburg's Ingenieurschule für Fahrzeugtechnik (Engineering School for Vehicle Technology), followed by a position as curator of the Deutsches Museum in Munich (1968–1977), where he was responsible for developing the land transport and petroleum/natural gas departments.

Since 1978, Erik Eckermann has been active as an automotive historian, with numerous books and contributions to magazines, radio, television, and educational courses to his credit, along with biographies of engineers, lectures, and appraisals. Further achievements have included planning and execution of automobile and vehicle exhibits, such as Auto Hall II at the Deutsches Museum (1985–1987); 750 Years of Traffic, Hannover (1991); 175 Years of Personal Transportation (1993); 100 Years of Trucks (1996); Wood in Automotive Construction (1997); 100 Years of Automotive Exhibitions (1997); and the Hanomag Technology Forum (2000). Most of these have been special exhibits at the Frankfurt International Auto Show and the Hannover Fair.

Mr. Eckermann's awards include the "Award of Distinction" from the Society of Automotive Historians (U.S.A.) for *Dynamik beherrschen* ("Mastering Dynamics"), a history of ITT-Teves, 1987; and The Deutsches Museum Prize for Publications for *Technikgeschichte im Deutschen Museum: Automobile*, 1989.